Communications in Computer and Information Science 1684

More information about this series at https://link.springer.com/bookseries/7899

Bernabé Dorronsoro · Mario Pavone ·
Amir Nakib · El-Ghazali Talbi (Eds.)

Optimization and Learning

5th International Conference, OLA 2022
Syracuse, Sicilia, Italy, July 18–20, 2022
Proceedings

Springer

Editors
Bernabé Dorronsoro ⓘ
Universidad de Cádiz
Cadiz, Spain

Mario Pavone ⓘ
University of Catania
Catania, Italy

Amir Nakib ⓘ
University of Paris-Est
Paris, France

El-Ghazali Talbi ⓘ
Université de Lille
Lille, France

ISSN 1865-0929   ISSN 1865-0937 (electronic)
Communications in Computer and Information Science
ISBN 978-3-031-22038-8   ISBN 978-3-031-22039-5 (eBook)
https://doi.org/10.1007/978-3-031-22039-5

This Springer imprint is published by the registered company Springer Nature Switzerland AG
The registered company address is: Gewerbestrasse 11, 6330 Cham, Switzerland

Preface

This book compiles the best submitted papers to the Fourth International Conference on Optimization and Learning (OLA 2022), which took place in Syracuse, Italy, from July 18 to July 20, 2022. The main objective of OLA 2022 was to attract influential researchers from all over the world in the fields of complex problems optimization, machine learning, and deep learning to discuss the synergies between these research fields and their applications for real-world problems, amongst other things. The conference aims to build a nice atmosphere where relevant researchers present their innovative works.

Three categories of papers were considered in OLA 2022, namely, ongoing research works, high impact journal publications (both in the shape of an extended abstract), and regular papers with novel contents and important contributions. A selection of the best papers in this latter category is published in this book.

The conference received a total of 52 papers, of which 46 were presented at OLA 2022 edition. The presentations were arranged into nine sessions, covering topics such as the use of optimization methods to enhance learning techniques, the use of learning techniques to improve the performance of optimization methods, advanced optimization methods and their applications, machine and deep learning techniques, or applications of learning and optimization tools to transportation and routing, scheduling, or other real-world problems.

The top 18 papers, according to review score, are included in this book, which is 30.8% of all submitted papers. All of the selected papers were reviewed in a single-blind process, with each paper receiving at least three reviews.

July 2022

Bernabé Dorronsoro
Mario Pavone
Amir Nakib
El-Ghazali Talbi

Organization

Conference Chairs

Mario Pavone   University of Catania, Italy
Lionel Amodeo   Université de Technologie de Troyes, France

Program Chairs

Bernabe Dorronsoro   University of Cadiz, Spain
Amir Nakib   University of Paris-Est Creteil, France

Steering Committee Chair

El-Ghazali Talbi   University Lille and Inria, France

Organization Committee

Jeremy Sadet   University Valenciennes, France
Rachid Ellaia   EMI and Mohamed V University of Rabat,
          Morocco
Rocco A. Scollo   University of Catania, Italy
Antonio M. Spampinato   University of Catania, Italy
Georgia Fargetta   University of Catania, Italy
Carolina Crespi   University of Catania, Italy

Publicity Chairs

Juan J. Durillo   Leibniz Supercomputing Center, Germany
Grégoire Danoy   University of Luxembourg, Luxembourg

Program Committee

Lionel Amodeo   Université de Technologie de Troyes, France
Mehmet-Emin Aydin   University of Bedfordshire, UK
Mathieu Balesdent   ONERA, France
Pascal Bouvry   University of Luxembourg, Luxembourg
Loïc Brevault   ONERA, France
Matthias R. Brust   University of Luxembourg, Luxembourg
Grégoire Danoy   Universiy of Luxembourg, Luxembourg

Contents

Optimization and Learning

Evolutionary-Based Co-optimization of DNN and Hardware Configurations on Edge GPU

Halima Bouzidi[1(✉)], Hamza Ouarnoughi[1], El-Ghazali Talbi[2],
Abdessamad Ait El Cadi[1], and Smail Niar[1]

[1] Université Polytechnique Hauts-de-France, LAMIH/CNRS, Valenciennes, France
Halima.Bouzidi@uphf.fr
[2] Université de Lille, CNRS/CRIStAL INRIA Lille Nord Europe, Lille, France

Abstract. The ever-increasing complexity of both Deep Neural Networks (DNN) and hardware accelerators has made the co-optimization of these domains extremely complex. Previous works typically focus on optimizing DNNs given a fixed hardware configuration or optimizing a specific hardware architecture given a fixed DNN model. Recently, the importance of the joint exploration of the two spaces draw more and more attention. Our work targets the co-optimization of DNN and hardware configurations on edge GPU accelerator. We investigate the importance of the joint exploration of DNN and edge GPU configurations. We propose an evolutionary-based co-optimization strategy for DNN by considering three metrics: DNN accuracy, execution latency, and power consumption. By combining the two search spaces, we have observed that we can explore more solutions and obtain a better tradeoff between DNN accuracy and hardware efficiency. Experimental results show that the co-optimization outperforms the optimization of DNN for fixed hardware configuration with up to 53% hardware efficiency gains for the same accuracy and latency.

1 Introduction and Related Works

Deep Neural Networks (DNN) and hardware accelerators are both leading forces for the observed progress in Machine Learning (ML). However, DNNs are becoming more and more complex and resource-demanding. Therefore, they need careful optimization to achieve the best tradeoff between accuracy and hardware efficiency. To meet this challenge, Hardware-aware Neural Architecture Search (HW-NAS) has been proposed [1] in which DNN hardware efficiency is considered during the exploration process. Nevertheless, hardware efficiency depends not only on the DNN architecture but also on the hardware configuration [2–4]. Most existing works on HW-NAS fall into the optimization of DNN without considering the reconfigurability of the hardware accelerator. As discussed in [5], this approach is sub-optimal because the HW-NAS search space is narrower when considering only a fixed hardware configuration. Thus, by considering the hardware design space, it is possible to find tailor-made DNNs for

B. Dorronsoro et al. (Eds.): OLA 2022, CCIS 1684, pp. 3–12, 2022.
https://doi.org/10.1007/978-3-031-22039-5_1

each hardware configuration and vice-versa. The joint exploration of both search spaces is referred to as *the co-optimization* in this paper.

Recent works [2, 6–13] have tackled the co-optimization problem where DNN architectures are optimized jointly with hardware configurations. Thereby, DNN-HW pairs are generated during the exploration process. However, as pointed out by [14], this strategy incurs a huge search time given the complexity of the joint search space. Therefore, another co-exploration strategy has been proposed by [15–17], in which separate optimization algorithms optimize DNN and hardware. The results can be then communicated between the two optimization algorithms to adjust the exploration process at some points. However, although this strategy solves the drawback of the first joint approach, the sub-optimality of the final results remains its critical issue. The works mentioned above can also be classified according to the following factors [5]: DNN search space and targeted hardware accelerator, exploration algorithm, objective functions, and fitness evaluation strategy. Nevertheless, only a few works have attempted to consider the co-optimization problem for GPU-based hardware platforms. Recent edge GPUs allow the reconfigurability of different hardware parameters such as processing units and operating frequencies. The impact of these parameters on DNN performance has been well discussed and analyzed in [18, 19]. Moreover, recent works shed light on the impact of these parameters when varying other parameters of the DNN. For instance, the authors in [3] adjust both the configuration of the GPU operating frequencies and the batch size of the DNN to maximize the inference hardware efficiency.

Our paper is structured as follows. In Sect. 2 we present and explain the motivation of our work. In Sect. 3 we first formulate our multi-objective co-optimization problem. Then, we describe and explain our optimization approach. Section 4, presents our experimental setup and results. Then we discuss our obtained results and compare them to other approaches and state-of-the-art solutions. Finally, the conclusion will be given in Sect. 5.

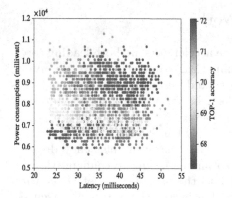

Fig. 1. The results of performing an HW-NAS under fixed edge GPU's hardware configuration.

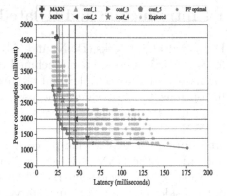

Fig. 2. The results of optimizing edge GPU's hardware configuration for a fixed DNN model.

2   Motivation

Figure 1 shows that different DNN models give different tradeoffs between accuracy and hardware efficiency (i.e., latency and power consumption) under a fixed hardware configuration. This figure gives the results of a Hardware-aware Neural Architectural Search (HW-NAS) [20] that we have performed under a fixed hardware configuration. As hardware platform, we used the NVIDIA Jetson AGX edge GPU. Each point represents a DNN model in the search space. The x-axis and y-axis represent the measured latency and power consumption, respectively, on the edge GPU. The color of the points indicates the TOP-1 accuracy of the DNN. Figure 1 shows that different tradeoffs are obtained between accuracy and hardware efficiency for each explored DNN model.

Figure 2 illustrates that the hardware efficiency of a single DNN varies when varying the hardware configuration. This figure gives the results of an exhaustive exploration of hardware configurations for a fixed DNN model, EfficientNet-B0 [21] in this case. To showcase the impact of the hardware configuration on the overall hardware efficiency of the DNN, we compare the obtained results with the predefined default hardware configurations proposed by NVIDIA. In this figure MAXN (resp. MINN) is the NVIDIA Jetson AGX configuration with the highest (resp. the smallest) allowed clock frequency. MAXN (resp. MINN) in general maximizes (resp. minimises) the processing speed at the cost of a high (resp. low) power consumption. The other configurations (i.e., from conf_1 to conf_5) are proposed to achieve a tradeoff between MAXN and MINN [22]. Remarkably, from the optimal Pareto front (marked in blue), the exhaustive exploration identifies hardware configurations that completely dominate all NVIDIA's predefined default configurations. It's important to mention that the Pareto front does not contain any configuration of the NVIDIA's predefined configurations. Furthermore, the dominant solutions in the Pareto front improve upon the default configurations of NVIDIA (i.e., MAXN and MINN) by 57% and 40% for power consumption and latency, respectively This result shows the necessity to explore the space of hardware configurations, to enhance the hardware efficiency of the DNN.

From these two figures we can conclude that, the performances of a DNN model are determined by the DNN architecture and the HW configuration. However, understanding the impact of these factors is not straightforward. For instance, the DNN with the highest computational complexity is not necessarily the most accurate model in the DNN search space [23]. In addition, the correlation between the performance metrics is also complex. For instance, minimizing latency incurs a maximization of power consumption and vice versa [24].

3   Proposed Approach

3.1   Problem Formulation

Our DNN-HW co-optimization problem can be formulated as a multi-objective problem. As exploring the whole HW space and the whole DNN space is time-consuming and costly in terms of development efforts, we need a tool for a rapid

DNN/HW co-design space exploration. Furthermore, while accelerating the co-design space exploration, the tool must adequately provide good approximation of the Pareto front. In this paper, we focus on solutions depicting the highest DNN accuracy and hardware efficiency. The term hardware efficiency refers to the trade-off between latency and power consumption. The mathematical formulation of our problem is as follows:

$$MOP : \begin{cases} \min F(dnn, hc) = [(Error_{dataset}(dnn), Latency_{HW_{ACC}}(dnn, hc), Power_{HW_{ACC}}(dnn, hc)]^T \\ s.t. \quad (dnn, hc) \in (DNN \times HWConf) \end{cases}$$

$$(1)$$

where dnn represents a DNN model defined by the DNN decision variables detailed in Table 1. hc represents a hardware configuration defined by the hardware decision variables listed in Table 1. DNN and $HWConf$ are the decision spaces of DNNs and hardware configurations, respectively, detailed in Table 1. Finally, F is the objective vector to optimize by minimizing the DNN error (i.e., maximizing accuracy) on $dataset$, DNN latency and power consumption on the hardware accelerator HW_{ACC}. We note that the $Error$ is measured by calculating the TOP-1 error rate, which is the percentage of images from $dataset$ for which the correct label is not the class label predicted by the DNN. $Latency$ is the execution time (in milliseconds) of dnn on the hardware accelerator HW_{ACC}. Finally, $Power$ is the average power consumption (in milliwatt) observed when executing dnn on the hardware accelerator HW_{ACC}.

3.2 Optimization Methodology

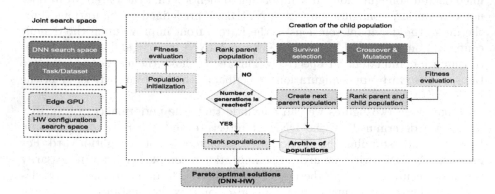

Fig. 3. Overview on the proposed co-optimization approach based on NSGA-II

To solve the above problem, we propose an evolutionary-based co-optimization strategy, where we search for both the optimal DNN architecture and hardware configuration. The search is done by exploring DNN and HW search spaces. Figure 3 details the proposed co-optimization approach. Our methodology includes three main components:

- **Joint Search Space:** We extend the search space of the HW-NAS by including the hardware configurations. Furthermore, by definition, the joint search space can be generalized to any DNN, task, dataset, and hardware accelerator. Thus, these four factors are considered as inputs in our co-optimization framework. In this paper, we use the joint search space detailed in Table 1. We note that all the considered decision variables are discrete.
- **Optimization Algorithm:** We choose NSGA-II [25] as an evolutionary algorithm to explore the joint search space. NSGA-II is a widely used algorithm for NAS problems in general, and HW-NAS in particular [26]. Moreover, it typically provides a fast and efficient convergence by searching a wide range of solutions. These two abilities is due to its selection strategy based on non-dominated sorting and crowding distance, which allow for both convergence and diversity of solutions. In this paper, the parameters used for NSGA-II are detailed in Table 2. We first initialize the population using the LHS method (Latin HyperCube Sampling). Then, next populations are generated from: 1) selecting the best solutions using the non-dominated sorting algorithm of NSGA-II and 2) applying mutation and crossover on these best solutions to create the offspring population. We choose a high crossover probability of 80% to increase the reproducibility of good candidate solutions. However, we decrease the probability of mutation to 30% to prevent the risk of losing traces of good candidate solutions. Crossover and mutation are chosen uniformly.
- **Evaluation Strategy:** The explored pairs are evaluated regarding the DNN accuracy and hardware efficiency. DNN accuracy is evaluated in two stages: 1) We use a fast evaluation technique to quickly determine the DNN accuracy during the exploration, then after the exploration, 2) we perform a more complete evaluation of the DNN accuracy for the elite solutions. We note that the results of the two evaluation techniques are highly correlated. Furthermore, DNN hardware efficiency is directly measured by executing the DNN on the hardware accelerator under the specified configuration.

4   Evaluation

Table 1. The joint search space of DNN and hardware parameters

Decision variables	DNN search space					Hardware search space		
	Input resolution	Width	Depth	Kernel size	Expand ratio	CPU frequency	GPU frequency	Memory frequency
Values	[192, 288]	[16, 1984]	[1, 8]	[3, 5]	[1, 6]	[0.1, 2.3]	[0.1, 1.4]	[0.2, 2.1]
Cardinality	4	16	8	2	4	29	14	9

Table 2. NSGA-II parameters

Parameter	Value
Number of generations	50
Population size	100
Population initialization	LHS
Mutation, probability	Uniform, 30%
Crossover, probability	Uniform, 80%

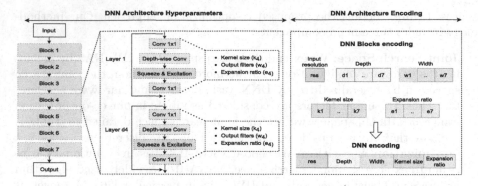

Fig. 4. DNN search space encoding: A candidate DNN architecture is real-encoded using a single vector that comprises five sub-vectors depicting: input resolution, depth, width, kernel size, and expansion ratio of each block

4.1   Experimental Setup

With the essential concepts described above, our co-optimization problem instance has the following inputs:

– *DNN search space:* We use the same search space provided by [20,27]. The search space contains 10^{11} DNN architectures, as detailed in Table 1 and Fig. 4. The authors in [20] provide a prediction model for accuracy assessment, as a fast evaluation tool, and a pretrained supernet, as a complete evaluation tool. However, the second strategy is time-consuming as the sampled DNN needs to be calibrated on the entire training dataset (ImageNet). Thus, we used a fast evaluation strategy during the exploration then performed a complete evaluation using the pretrained supernet for the elite solutions.
– *Dataset:* We choose to explore the joint search space for a state-of-art dataset such as ImageNet. Thus, the DNN accuracy are calculated on the ImageNet [28] validation dataset. All images are preprocessed using data augmentation techniques such as whitening, upsampling, random cropping, and random horizontal flipping, before feeding them to the DNN.
– *Hardware search space:* We choose the NVIDIA Jetson AGX Xavier GPU as a hardware accelerator [29]. NVIDIA Jetson GPU accelerators allow the reconfigurability of different hardware parameters such as the number of operating cores. It also allows to have different operating clock frequency in the cores, GPU, and memory units. The chosen values of these parameters depend on the application requirements. For our case, we only vary the operating frequencies as detailed in Table 1 (Fig. 9).

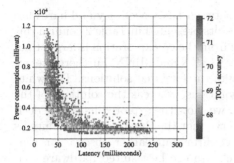

Fig. 5. Co-exploration results on the joint search space

Fig. 6. Results of the three exploration approaches

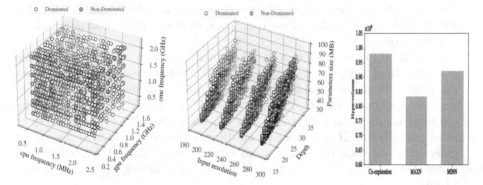

Fig. 7. Explored hardware configurations

Fig. 8. Explored DNN models

Fig. 9. Hypervolume results of the three optimization approaches

4.2 Experimental Results

In this section we will discuss the obtained results from two main perspectives:

- *Efficiency of the co-exploration:* To underline the co-exploration's importance, we compare the results obtained when co-exploring the joint search space and when performing a typical HW-NAS under fixed hardware configurations. We choose two default configurations proposed by the hardware manufacturer, NVIDIA in our case: MAXN and MINN. Figure 5 gives the results of the co-exploration, where Fig. 6 depicts a comparison between the results of the three approaches (i.e., joint, MAXN, and MINN), marked with different points shapes. After analyzing the two figures, we can clearly see that the region explored in the joint space is much larger than the regions explored when fixing the hardware configuration to MAXN or MINN. Furthermore, the explored regions when fixing the hardware configuration are included in the co-exploration. Indeed, the joint search space allows for exploring much larger solutions and hence different tradeoffs between DNN accuracy and hardware efficiency. The obtained hypervolume results presented in Table 2 confirm this observation. The obtained hypervolume from the co-exploration is larger than the hypervolumes of the HW-NAS under MAXN and MINN. Furthermore, we give Figs. 7 and 8 to show the diversity of the explored solutions. The white points correspond to all the explored solutions, whereas the solutions of the Pareto set are marked in blue. In Fig. 8, we show the explored hardware decision space. From this figure, we can observe that the Pareto optimal solutions are diverse and well distributed. This confirms our earlier observation that no a priori knowledge can be used to choose the best-suited hardware configuration without actual exploration and evaluation. Figure 7 gives the characteristics of the explored DNN models in terms of input resolution, depth (i.e., number of layers), and size of trainable parameters (in Mega-Bytes). Similarly, the Pareto optimal solutions are well distributed and diverse. This also supports the importance of the exploration as we can assume a priori which DNN model will be Pareto optimal without actual evaluation of its performance.

- *Optimality of the obtained results:* To further investigate the efficiency of the co-exploration, we select top pairs of (DNN, hardware configuration) from the Pareto front and compare them to SOTA DNN models under the widely used default configuration proposed by NVIDIA, MAXN. Table 1 details the obtained results. Our co-optimization approach was able to identify better solutions in terms of accuracy and hardware efficiency. We can notice power gains of up to 53% under the same latency constraints. Furthermore, we observe an accuracy improvement of up to 0.5% on the ImageNet dataset (Table 3).

Table 3. Performance of the baseline models proposed by AttentiveNAS [20] compared to our top solutions of (DNN,hw-conf) obtained from the Pareto front approximation of the co-optimization

DNN, hw_conf	TOP-1 acc (%)	Latency (ms)	Power consumption (mw)
AttentiveNAS-A2, MAXN	78.8	29.91	6575
AttentiveNAS-A3, MAXN	79.1	33.51	6575
AttentiveNAS-A4, MAXN	79.8	32.67	7033
AttentiveNAS-A5, AMXN	80.1	35.66	6881
Ours-B0, hc0	**79.0**	**28.85**	**4744**
Ours-B1, hc1	**79.6**	**30.82**	**4591**
Ours-B2, hc2	**79.9**	**33.03**	**4591**
Ours-B3, hc3	**80.2**	**34.10**	**6118**

5 Conclusion

In this paper, we investigated the importance of the joint exploration of DNN and hardware configurations for edge GPU accelerators. We propose a co-optimization approach based on an evolutionary algorithm (NSGA-II) to explore these two search spaces. We aim was to minimize three objective functions: DNN TOP-1 error, latency, and power consumption. Experimental results on the Jetson AGX Xavier demonstrated the efficiency of the co-optimization compared to typical HW-NAS under fixed hardware configurations. Moreover, the top pairs found by our co-optimization are more energy-efficient with up to 53% gains than solutions found by state-of-the-art models under the same accuracy and latency constraints. As future work, we plan to enhance our co-optimization strategy by proposing more efficient selection and recombination operators for the optimization algorithm. We also aim to investigate more hardware configurations and DNN benchmarks to showcase the importance of co-optimization.

References

1. Benmeziane, H., El Maghraoui, K., Ouarnoughi, H., Niar, S., Wistuba, M., Wang, N.: A comprehensive survey on hardware-aware neural architecture search. CoRR, abs/2101.09336 (2021)
2. Hao, C., et al.: FPGA/DNN co-design: an efficient design methodology for 1ot intelligence on the edge. In: 2019 56th ACM/IEEE Design Automation Conference (DAC), pp. 1–6. IEEE (2019)
3. Nabavinejad, S.M., Reda, S., Ebrahimi, M.: Coordinated batching and DVFS for DNN inference on GPU accelerators. IEEE Trans. Parallel Distrib. Syst. (2022)
4. Yang, L., et al.: Co-exploration of neural architectures and heterogeneous ASIC accelerator designs targeting multiple tasks. In: 2020 57th ACM/IEEE Design Automation Conference (DAC), pp. 1–6. IEEE (2020)
5. Jiang, W., et al.: Hardware/software co-exploration of neural architectures. IEEE Trans. Comput.-Aided Design Integr. Circ. Syst. **39**(12), 4805–4815 (2020)
6. Lin, Y., Hafdi, D., Wang, K., Liu, Z., Han, S.: Neural-hardware architecture search. In: NeurIPS WS (2019)
7. Li, Y., et al.: EDD: efficient differentiable DNN architecture and implementation co-search for embedded AI solutions. In: 2020 57th ACM/IEEE Design Automation Conference (DAC), pp. 1–6. IEEE (2020)
8. Jiang, W., Yang, L., Dasgupta, S., Jingtong, H., Shi, Y.: Standing on the shoulders of giants: hardware and neural architecture co-search with hot start. IEEE Trans. Comput. Aided Des. Integr. Circ. Syst. **39**(11), 4154–4165 (2020)
9. Chen, W., Wang, Y., Yang, S., Liu, C., Zhang, L.: You only search once: a fast automation framework for single-stage DNN/accelerator co-design. In: 2020 Design, Automation & Test in Europe Conference & Exhibition (DATE), pp. 1283–1286. IEEE (2020)
10. Choi, K., Hong, D., Yoon, H., Yu, J., Kim, Y., Lee, J.: DANCE: differentiable accelerator/network co-exploration. In: 2021 58th ACM/IEEE Design Automation Conference (DAC), pp. 337–342. IEEE (2021)
11. Liang, Y., et al.: An efficient hardware design for accelerating sparse CNNs with NAS-based models. IEEE Transactions on Computer-Aided Design of Integrated Circuits and Systems **41**, 597–613 (2021)

12. Zhou, Y., et al.: Rethinking co-design of neural architectures and hardware accelerators. arXiv preprint arXiv:2102.08619 (2021)
13. Pinos, M., Mrazek, V., Sekanina, L.: Evolutionary neural architecture search supporting approximate multipliers. In: Hu, T., Lourenço, N., Medvet, E. (eds.) EuroGP 2021. LNCS, vol. 12691, pp. 82–97. Springer, Cham (2021). https://doi.org/10.1007/978-3-030-72812-0_6
14. Sekanina, L.: Neural architecture search and hardware accelerator co-search: a survey. IEEE Access 9, 151337–151362 (2021)
15. Lu, Q., Jiang, W., Xu, X., Shi, Y., Hu, J.: On neural architecture search for resource-constrained hardware platforms. arXiv preprint arXiv:1911.00105 (2019)
16. Abdelfattah, M.S., Dudziak, Ł., Chau, T., Lee, R., Kim, H., Lane, N.D.: Best of both worlds: AutoML codesign of a CNN and its hardware accelerator. In: 2020 57th ACM/IEEE Design Automation Conference (DAC), pp. 1–6. IEEE (2020)
17. Cai, H., Gan, C., Wang, T., Zhang, Z., Han, S.: Once-for-all: train one network and specialize it for efficient deployment. arXiv preprint arXiv:1908.09791 (2019)
18. Spantidi, O., Galanis, I., Anagnostopoulos, I.: Frequency-based power efficiency improvement of CNNs on heterogeneous IoT computing systems. In: 2020 IEEE 6th World Forum on Internet of Things (WF-IoT), pp. 1–6. IEEE (2020)
19. Liu, S., Karanth, A.: Dynamic voltage and frequency scaling to improve energy-efficiency of hardware accelerators. In: 2021 IEEE 28th International Conference on High Performance Computing, Data, and Analytics (HiPC), pp. 232–241. IEEE (2021)
20. Wang, D., Li, M., Gong, C., Chandra, V.: AttentiveNAS: improving neural architecture search via attentive sampling. In Proceedings of the IEEE/CVF Conference on Computer Vision and Pattern Recognition, pp. 6418–6427 (2021)
21. Tan, M., Le, Q.: EfficientNet: rethinking model scaling for convolutional neural networks. In: International Conference on Machine Learning, pp. 6105–6114. PMLR (2019)
22. Jetson developer kits and modules. https://docs.nvidia.com/jetson/l4t/. Accessed 01 May 2021
23. Bianco, S., Cadene, R., Celona, L., Napoletano, P.: Benchmark analysis of representative deep neural network architectures. IEEE Access 6, 64270–64277 (2018)
24. Tang, Z., Wang, Y., Wang, Q., Chu, X.: The impact of GPU DVFs on the energy and performance of deep learning: an empirical study. In: 10th ACM International Conference on Future Energy Systems, pp. 315–325 (2019)
25. Deb, K., Agrawal, S., Pratap, A., Meyarivan, T.: A fast elitist non-dominated sorting genetic algorithm for multi-objective optimization: NSGA-II. In: Schoenauer, M., et al. (eds.) PPSN 2000. LNCS, vol. 1917, pp. 849–858. Springer, Heidelberg (2000). https://doi.org/10.1007/3-540-45356-3_83
26. Liu, Y., Sun, Y., Xue, B., Zhang, M., Yen, G.G., Tan, K.C.: A survey on evolutionary neural architecture search. IEEE Trans. Neural Netw. Learn. Syst. (2021)
27. Wang, D., Gong, C., Li, M., Liu, Q., Chandra, V.: AlphaNet: improved training of supernets with alpha-divergence. In: International Conference on Machine Learning, pages 10760–10771. PMLR (2021)
28. Krizhevsky, A., Sutskever, I., Hinton, G.E.: ImageNet classification with deep convolutional neural networks. In: Pereira, F., Burges, C.J.C., Bottou, L., Weinberger, K.Q. (eds.) Advances in Neural Information Processing Systems, vol. 25. Curran Associates Inc. (2012)
29. Jetson AGX Xavier developer kit. https://developer.nvidia.com/embedded/jetson-agx-xavier-developer-kit. Accessed 01 Feb 2021

Maximum Information Coverage and Monitoring Path Planning with Unmanned Surface Vehicles Using Deep Reinforcement Learning

Samuel Yanes Luis[✉], Daniel Gutiérrez Reina, and Sergio Toral

University of Seville, Sevilla, Spain
{syanes,dgutierrezreina,storal}@us.es

Abstract. Manual monitoring large water reservoirs is a complex and high-cost task that requires many human resources. By using Autonomous Surface Vehicles, informative missions for modeling and supervising can be performed efficiently. Given a model of the uncertainty of the measurements, the minimization of entropy is proven to be a suitable criterion to find a surrogate model of the contamination map, also with complete coverage pathplanning. This work uses Proximal Policy Optimization, a Deep Reinforcement Learning algorithm, to find a suitable policy that solves this maximum information coverage path planning, whereas the obstacles are avoided. The results show that the proposed framework outperforms other methods in the literature by 32% in entropy minimization and by 63% in model accuracy.

Keywords: Deep reinforcement learning · Informative path planning · Autonomous surface vehicles

1 Introduction

More than 80% of human activities wastewater is discharged into rivers and seas without prior treatment, making the task of monitoring hydrological resources essential for the sustainability of the planet and developing populations [1]. Manual monitoring of water quality in very large lakes and rivers is a costly task, especially when the environment is polluted and can pose a risk to field operators. Additionally, manual monitoring is inefficient, as manned vessels are heavier and tend to use fossil fuels. On the other hand, the deployment of static sensor networks has certain disadvantages. The measurement points are fixed and cannot change their trajectory depending on the information needs of biologists and authorities.

This work proposes the use of autonomous surface vehicles (ASVs) for dynamic monitoring of biologically at risk scenarios, such as Lake Ypacaraí,

S. Y. Luis—Participation financed by the Ministry of Universities under the FPU-2020 grant of Samuel Yanes Luis and by the Regional Govt. of Andalusia under PAIDI 2020 funds - P18-TP-1520.

where spills and uncontrolled eutrophication have caused a dangerous bloom of blue-green algae colonies. With these electric vehicles equipped with high-quality sensor modules to measure turbidity, ammonium, dissolved oxygen, etc., multi-objective monitoring missions can be carried out (see Fig. 1). The use of these vehicles, however, requires the development of an intelligent routing module capable of dealing with the environmental monitoring requirements, which are: planning obstacle-free paths, sampling the entire objective surface with minimization of the redundancy, and a surrogate model that truly represents the sampled variables. The informativeness criterion, from the perspective of Information Theory, can be designed using entropy as an indicator of the model uncertainty of the lake variables. Let there be a surrogate model $\hat{f}(X)$, representing the value of a pollution variable in the navigable domain $X \in \mathbb{R}^2$, the reduction of the entropy $H(X|f)$ by incorporating new information implies a decrease of the information clutter, that is, the certainty about the obtained model. The objective of an information planner will then be to maximize the information gain ΔI from an instant t to $t+1$.

Finding the set of samples that reduce the entropy of the process while avoiding the nonnavigable areas makes this problem, called the Maximum Information Coverage Path Planning Problem (MICPP), nonpolynomial hard. Due to the unmanageable dimension of the possibilities in tracing a route within the lake that meets the requirements, it is necessary to use Artificial Intelligence (AI) or Metaheuristic Optimization techniques to find solutions that, in most cases, are suboptimal. In this work, we propose the use of deep reinforcement learning (DRL) algorithms to find solutions to the problem. With an a priori model of the covariance of the variables (kernel) and the use of a robust optimization algorithm of a deep policy (PPO), a framework is proposed that makes it possible to find good solutions to the MICPP problem in a reasonable amount of time. Furthermore, this framework allows one to obtain routes that meet a second objective: detecting possible pollution peaks that are difficult to model. As the treatment of the information is independent of any physical model and only bases its optimality on the entropy decrease, it can be applied in a wide range of cases: gas leakage detection, electromagnetic indoors characterization, patrolling and surveillance, etc.

Fig. 1. ASV prototype (right) developed for monitoring the Lake Ypacaraí. The ASV is equipped with a high performance sensor module that is able to measure pH, dissolved oxygen, turbidity, and ammonia concentration. Every mission starts from the deploy zone marked in the map (left).

The main contributions of this work are:

- A Deep Reinforcement Framework to solve the Maximal Distributed Information Coverage problem and global path planning in the Ypacaraí Lake.
- An analysis of the stability, generalization, and performance of the proposed DRL approach and other well-known planning heuristics.

This article is organized as follows. In Sect. 2, an overview of recent advances in patrolling and monitoring using autonomous vehicles is presented. In Sect. 3, the problem is presented formally with its assumptions, and the DRL framework is explained. In Sect. 4, the simulations are described with an analysis of the results. Finally, the conclusions and future lines are described in Sect. 5.

2   State of the Art

The use of Unmanned Surface Vehicles for monitoring aquatic environments is becoming increasingly common thanks to the development of robotics technology and battery autonomy. In [2], the use of inexpensive surface vehicle swarms is proposed for sea border patrolling and environmental monitoring. In [3], an aquatic robot is also used to perform an autonomous bathymetry study in oceanographic contexts. These works use sensor modules and network connectivity to perform their tasks and avoid obstacles.

If we focus on the task of monitoring hydrological resources, we can separate the contributions of the literature into three topics: modeling [4,5], patrolling [6,7], and coverage [8,9]. In the first branch, the ultimate goal is to find a scalar field that represents the environmental variables of water (turbidity, pH, etc.). In works such as [4], the use of Bayesian optimization algorithms together with Gaussian processes is proposed for obtaining faithful models with few samples. In this sense, our proposal includes the use of Gaussian processes as a way to obtain a model of the environment. Other contributions such as [5] explicitly work on multi-objective optimization, while in our proposal the coverage tries to find a path independent of the underlying variables. In the second application, ASVs are typically used to solve the homogeneous [6,7] and heterogeneous patrolling problem. In the first case, the aim is to find periodic routes that minimize the average visit time of each area, while in the second case, this time is weighted according to a previously specified importance map. These approaches have in common that the map is modeled as a discrete, metric, undirected graph, the resolution of which severely affects the dimension of the problem. In this new framework, on the other hand, the resolution of the map does not affect the complexity of the problem, since the state and action space are continuous. In the third application, vehicles can be used to cover, given a path length, the maximum possible area [8]. This work proposes to maximize the area covered by the vehicle by using a Genetic Algorithm that penalizes passing through already visited areas. In [9], a similar approach is used to adapt to the detection of cyanobacterial blooms in the same Lake Ypacaraí. Both approaches to the full coverage problem have as a counterpart that the vehicle is limited to perform straight trajectories from shore to shore of the lake. In addition, coverage is considered binary (covered or uncovered). Both limitations are overcome in this work when it is specified, in the first case, that the action space can take any direction at each instant and, in the second case, that the coverage level is measured with a smooth function in the circular surrounding of every sample (radial kernel), since the underlying real scalar field can be considered smooth as well.

Regarding the use of DRL for autonomous vehicle monitoring, there are many examples in the literature that implement deep policy optimization. In the case of [10], a Monte Carlo optimization is used to minimize entropy in the task of monitoring crop fields. The main differences of our proposal with this one are i) the action space used in this work is continuous, which increases the complexity of the solutions, and ii) in [10] terrain constraints are not taken into account. In this work, we consider that the agent must adapt to a real morphology with non-navigable areas, which means not only finding informative routes, but also avoiding obstacles.

3   Methodology

3.1   Sequential Decision Problem

The MICPP simultaneously addresses two learning problems: i) the vehicle must find navigable routes as long as possible to maximize the sampled area, and ii) these routes must optimally minimize the information entropy given a model of uncertainty of environmental variables. This is a sequential decision problem, since the vehicle must choose, at each instant, which next point p^{t+1} to move to achieve both objectives. This sequential process is defined as a Markov Decision Process (MDP) in which the scenario with state s_t, the agent performs an action a_t according to a policy $\pi(s_t)$ that maps s to a which generates a reward r_t according to a reward function $R(s_t, a_t)$. Within this framework, the ASV must learn to decide at every instant the next point of movement that maximizes the episodic discounted reward. The optimal policy $\pi^*(s_t)$ is:

$$\pi^* = \max_{\pi} \sum_{t=0}^{T} R^{\pi}(s_t, a_t) \tag{1}$$

In this mathematical context, the ASV takes a sample of the water variables, updates its surrogate model and uncertainty, and decides which point to move to until a complete path is completed. In this work, we model the uncertainty of the process using a decreasing radial function (RBF kernel). This kernel function maps the correlation of two samples (x, x') depending on the Eulerian distance d and a scaling parameter l. The further the physical points, the lower the correlation between the samples.

$$K(x, x') = exp\left(-\frac{d(x, x')^2}{2l}\right) \tag{2}$$

This a priori model of the correlation is selected because it fits the only hypothesis that the underlying ground truth is smooth and the data behave like a multivariate Gaussian distribution (MGD). In this way, it is possible to obtain the correlation between a set of observation points X_{obs} by evaluating pairwise every physical point on the map:

$$\Sigma[X] = K(x, x') \quad \forall x, x' \in X \tag{3}$$

Hence $\Sigma[X_{obs}]$, constitutes the covariance matrix of the observation points. When a new measurement is incorporated into the measured set $X[meas]$, the conditioned correlation matrix noted as $\Sigma[X_{obs}|X_{meas}]$ can be calculated as:

$$\Sigma[X_{obs}|X_{meas}] = \Sigma[X_{obs}] - \Sigma[X_{obs}, X_{meas}] \times \Sigma[X_{meas}]^{-1} \times \Sigma[X_{meas}, X_{obs}] \tag{4}$$

This conditioned correlation indicates the uncertainty of the map given an equally distributed set of observation points and the new measuring points as in Fig. 2.

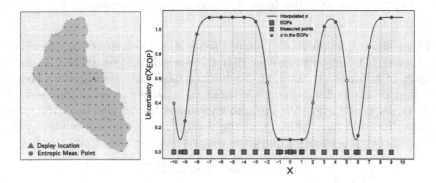

Fig. 2. Process of conditioning. The green squares represent the observation point for the entropy measurement. Sampling (red crosses) diminishes the uncertainty in those points according to Eq. (4), as the conditioned covariance between X_{obs} and X_{meas} decreases. (Color figure online)

Then, with the MGD hypothesis, the conditioned entropy $H[X_{obs}|X_{meas}]$ and the information gain are defined as

$$H|\Sigma[X_{obs}|X_{meas}]| = \frac{1}{2}\log(\Sigma[X_{obs}|X_{meas}]) + \frac{D}{2}(1 + \log(2\pi)) \tag{5}$$

$$\Delta I(X^t|a) = H(X^t) - H(X^{t+1}|a) \tag{6}$$

with D as the dimension of $\Sigma[X_{obs}|X_{meas}]$ [11].

Once the path is completed and different samples are obtained at each visited point, this information can be used to obtain a model of contamination of the environment. Since the model is obtained at the end of the collection of maximally distributed points, the regression method is independent of decision-making.

With respect to the action space of the agent A, it is defined to be continuous $A \in [-1, 1]$. An action a_t represents the heading angle $\psi \in [-\pi, \pi]$ of the straight trajectory between the next point and the current one, if possible, with a constant distance between them of d_{meas} (see Fig. 3). The ASV is restricted to a distance budget of 40 km due to the capacity of the battery at a constant speed of 2 m/s. At every reached point, a measurement is taken, and the model is updated. The path is considered complete when the maximum distance $D_{max} = 40$ km is reached.

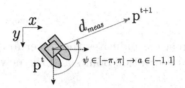

Fig. 3. Movement of the drone. The action space is contained in $[-1, 1]$.

The reward function $R(s, a)$ should sequentially represent the objective to be achieved. It is logical to impose that the reward function should be directly the gain of information $\Delta I(s, a)$. To further introduce the objective that the ASV should have collision-free trajectories, a penalty $c = -1$ is imposed when the next point p^{t+1} chosen by the policy cannot be reached from the current position. To avoid large changes in the reward range, the information gain is bounded between -1 and 1. Thus, the reward function would be:

$$R(s_t, a_t) = \begin{cases} min(max(\Delta I(s_t, a_t), -1), 1) & \text{if valid.} \\ c & \text{otherwise} \end{cases} \tag{7}$$

3.2 DRL Framework

To learn a policy, the Proximal Policy Optimization (PPO) algorithm is chosen, which is an on-policy reinforcement learning algorithm. In PPO, a deep policy $\pi(s|\theta)$ is updated using the stochastic gradient descent (SGD) approach over a loss function. This loss evaluates the advantage of each action in each state as a function of its reward and weights the gradient step depending on the difference between the policy prior to the optimization step and the new optimized policy. By limiting the distance between them, either by saturating the step by ϵ or penalizing the Kullback-Leibler (KL) distance between them, it is possible to robustly optimize the behavior without incurring instability so easily.

$$L(s, a, \theta_k, \theta) = min \left[\frac{\pi_\theta(a \mid s)}{\pi_{\theta_k}(a \mid s)} A^{\pi_{\theta_k}}(s, a) \text{clip} \left(\frac{\pi_\theta(a \mid s)}{\pi_{\theta_k}(a \mid s)}, 1 - \epsilon, 1 + \epsilon \right) A^{\pi_{\theta_k}}(s, a) \right]$$
$$\tag{8}$$

In this paper, the double constraint on the policy update has been imposed to make training more robust: both $KL(\pi_{\theta_k}, \pi_\theta)$ and the fraction in (7) are penalized. In the PPO, a two-headed neural network is trained. The first head corresponds to the value function $V(s)$ used in the advantage value $A^{\pi_{\theta_k}}(s, a)$, trained with the accumulated discounted reward R after the n steps of every mission. The other head directly returns the action bounded by the action limit $[-1, 1]$.

The state s_t for the PPO is defined to contain all information available in the problem. In this way, s_t is as a 3-channel image of $H \times W$ pixels. These channels correspond to the following. I) A binary representation of the navigation map $N_{map} \in \mathbb{R}$. II) The standard deviation σ of each physical point on the map of the RBF kernel, evaluated in the form of Eq. 2 and represented with the image shape. III) The discretized path that the ASV has completed so far, represented by pixels with values in $[0, 1]$, where 1 corresponds to the actual position and 0 for the starting position. The latter channel merges temporal dependencies to overcome the Markovian assumption of the reward: it must only depend on the current state and the current action.

Regarding the deep policy, the Convolutional Neural Network (CNN) is implemented for estimating the features of the state and processing the next action. The CNN backbone is composed of four 2D convolutional layers of [128, 64, 16, 16] filters, respectively. After the convolutional structure, a dense neural network (DNN) transforms the features into an action $\pi(s)$ and a state-value $V(s)$. The DNN is composed of 3 lineal layers of 256 neurons and 3 layers of 64 neurons in every separate head for the action and state-value. The activation function is the Leaky Rectifier Linear Unit (Leaky-ReLu) to avoid the dying-ReLu effect of low-value gradients. See Fig. 4 for the complete architecture.

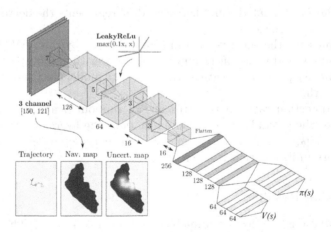

Fig. 4. CNN proposed for the deep policy. It is composed by a convolutional backbone of 4 blocks and a two-head dense block. The activation function is Leaky-ReLu.

4   Simulations and Results

This section discusses the hyperparameterization of the PPO algorithm, the usefulness of the reward function in terms of stability, and finally, the comparison of the results with other algorithms proposed in the literature for similar problems. To compare the model accuracy, a Shekel function[1] with randomly positioned and random size peaks is implemented to represent the scalar contamination field f to cover (see Fig. 5). The simulations were executed using an AMD Ryzen9 3900 (3.8 GHz) with a Nvidia RTX 2080 Super-8 GB GPU and 16 GB of RAM.

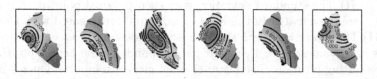

Fig. 5. Six random ground truths representing different scalar variables that the ASV must cover and model in the informative mission. The generator function is the Shekel function with a random number of peaks with random sizes uniformly distributed across the map.

4.1   Evaluation Metrics

For the analysis of the results, different metrics have been used to describe the utility of the learned policy.

– **AER:** The Accumulated Episodic Reward. It represents the decrease in total entropy over the mission time.
– **A^{nr}:** Represents the effective area in (km^2) covered by the ASV. A zone x is considered covered if the uncertainty $\sigma(x)$ is lower than 0.05.
– **MSE:** Mean square error between the surrogate model used and the variables ground truth.
– **ξ:** Peak detection rate. In the presence of random peaks k, that is, local maxima of the ground truth of contamination, the average rate of detected peaks $\xi = \mathbb{E}[k_{detected}/k]$. A local maximum is considered detected when the uncertainty in its location is lower than 0.05.

4.2   Learning Results

For the simulations, the hyperparameters in Table 1 were used. The PPO algorithm is less sensible to the selection of hyperparameters, and the following values were selected by trial. Every execution is equivalent to 1×10^6 steps. The learning rate was linearly annealed from 1×10^{-4} to 1×10^{-5} to enhance the stability

---

[1] https://deap.readthedocs.io/en/master/code/benchmarks/shekel.py.

of the learning and avoid catastrophic forgetting. In regards to the termination condition of the episode, two different approaches have been tested: apply the penalization and end the episode when colliding, or apply the penalization and let the episode continue from the same previous state.

Table 1. Hyperparameters involved in the PPO algorithm.

Hyperparameter	Value
Clip value (ϵ)	0.2
Learning rate (α)	$1 \times 10^{-4} \rightarrow 1 \times 10^{-5}$
max. KL permitted	5
Discount factor (γ)	0.95
Batch size ($\mathcal{B}$)	64
Entropy loss regularization	0.01
Surrogate loss horizon ($n_s teps$)	512
Collision penalty (c)	-1
Kernel lengthscale (l)	10

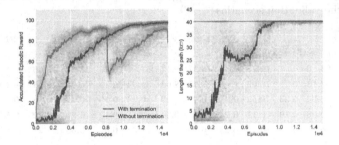

Fig. 6. Learning curves.

In Fig. 6 is shown the learning curves of the optimization process. The optimization shows a robust convergence to a high-reward solution. It is observed that the results of the first variant of the algorithm are more robust to the typical learning instabilities of DRL. The continuous penalty in the case of failure to terminate the episode degenerates the policy to the point of unlearning. The two main tasks can be considered learned: the length of the episode grows monotonically on average throughout the process until the maximum path length of 40 km is reached. This indicates that the deep agent assimilates the terrain constraints. On the other hand, the task of entropy minimization is effectively performed, since the trajectories have an increasingly informative character as the training proceeds.

It is important to note that the policy resulting from optimization with PPO $\pi(s; \theta)$ is a stochastic policy whose output is a Gaussian distribution $\mathcal{N}(\mu, \sigma)$. As training progresses, $pi(s; \theta)$ becomes more deterministic with a smaller standard deviation. This results in the fact that, despite having learned to synthesize collision-free trajectories, it happens that sampling an action may produce one. To verify the best behavior learned, the performance evaluation must force the action with the highest probability $a_t = median\,[\pi(s, \theta)]$.

4.3   Metric Comparison with Other Methods

To compare the results of the proposed method, five different heuristics from the literature have been implemented: i) a lawn mower (LM) trajectory, ii) GA optimization, iii) a collision-free random search, iv) greedy policy with respect to $\sigma(X)$, and v) greedy policy with respect to the expected improvement of a Gaussian process like in [4]. Ten different scenarios not seen in DRL optimization have served as a validation pool. In relation to MSE metrics, two regression methods have been used with the samples taken: the first is a Gaussian process as in [4], and the second method is a Support Vector Regressor (SVR). Both regression models use an RBF kernel with the same parameters as the one proposed to compute the uncertainty model.

Table 2. Statistical results of different approaches for the coverage and informative coverage in the validation scenarios-

Metric	LM		GA		σ-greedy		EI-greedy		Random		DRL	
	μ	σ	μ	σ	μ	σ	μ	σ	μ	σ	μ	σ
AER	83.10	0	72.75	0	70.31	9.17	57.28	9.19	53.61	10.36	**98.75**	7.08
A^{nr}	62.13	0	70.09	0	57.44	8.03	39.74	6.15	46.56	9.94	**94.91**	8.74
ξ		0.54	0.23	0.52	0.28	0.48	0.29	0.34	0.15	0.42	0.18	**0.72**	0.20
MSE_{gp}		0.08	0.10	0.11	0.13	0.05	0.04	0.18	0.13	0.20	0.22	**0.04**	0.09
MSE_{svr}		0.25	0.12	0.35	0.23	0.30	0.11	0.54	0.67	0.55	0.23	**0.07**	0.03

In Table 2, the comparative results are presented. It can be seen that the proposed algorithm is able to realize much more spatially distributed trajectories for entropy minimization. The entropy minimization factor reflected in the AER and the effective coverage area is significantly higher in the proposed algorithm (32% higher on average with the other approaches). It is logical to think that these two metrics are closely related since decreasing entropy leads to visiting uncovered areas and vice versa. In Fig. 7, the statistical results of the execution of the different approaches in the validation set are depicted. The proposed approach not only obtains better results but also faster within a mission time objective. This is related to the detection rate of singular events xi. Deep policy, as a direct consequence of generating a highly distributed path with low redundancy, achieves 33% more detections than the best-positioned algorithm

(LM). The LM algorithm generates a very intensive path, which achieves good homogeneous coverage, but too intensive due to monitoring redundancy.

In the MSE using different regression methods, we obtain very good values with the PPO algorithm, with an improvement of up to 67% using a GPR. This metric is greatly affected by information redundancy: visiting already covered areas does not provide extra information, so the proposed algorithm, through entropy reduction, manages to clearly beat any other informative trajectory for a wide range of possible ground truths. The fact that this is the case whether using a GP or an SVR also indicates that entropy is a good criterion for point selection in scalar-field regression.

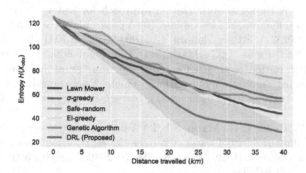

Fig. 7. Process entropy $H(X_{obs})$ over a 40 km monitoring mission.

5   Conclusions

In this paper, a new framework based on Deep Reinforcement Learning has been proposed for minimizing entropy in environmental scenarios using autonomous vehicles. Training by the PPO algorithm using convolutional networks and a graphical formulation of the state results in informative paths with high information utility. The proposal is capable of beating other algorithms and heuristics in maximizing information gain, which leads to improvements in other related metrics more or less directly related to entropy: area covered, location of singular events, and decreasing error in an arbitrary regression model. The latter is of interest because the algorithm does not depend on a particular type of model to work, but rather the improvement follows as a consequence of the information criterion. Furthermore, this approach raises the possibility of training on scenarios with arbitrary boundary conditions and simultaneously solving the task of obstacle avoidance and informational patrolling. It is proposed that, in future lines of work, a temporal factor can be included in the coverage for the entropy to be reduced on the time axis also for nonstationary environments. Furthermore, a next step is the application of different importance criteria, the decrease of entropy: a non-homogeneous informative coverage that emphasizes areas of high contamination.

References

1. United Nations Organization: United Nations Organization. Objective 6 (2021). https://www.un.org/sustainabledevelopment/es/water-and-sanitation/. Accessed 12 Nov 2021
2. Velez, F.J., et al.: Wireless sensor and networking technologies for swarms of aquatic surface drones. In: 2015 IEEE 82nd Vehicular Technology Conference (VTC2015-Fall), pp. 1–2 (2015). https://doi.org/10.1109/VTCFall.2015.7391193
3. Ferreira, H., et al.: Autonomous bathymetry for risk assessment with ROAZ robotic surface vehicle. OCEANS 2009-EUROPE, pp. 1–6 (2009). https://doi.org/10.1109/OCEANSE.2009.5278235
4. Peralt Samaniego, F., Gutierrez-Reina, D., Toral Marin, S.L., Arzamendia, M., Gregor, D.O.: A Bayesian optimization approach for water resources monitoring through an autonomous surface vehicle: the Ypacarai lake case study. IEEE Access **9**, 9163–9179 (2021). https://doi.org/10.1109/ACCESS.2021.3050934
5. Samaniego, F.P., Reina, D.G., Marin, S.L.T., Arzamendia, M., Gregor, D.O.: A Bayesian optimization approach for multi-function estimation for environmental monitoring using an autonomous surface vehicle: Ypacarai lake case study. Electron. (Switz.) **10**(963), 9163–9179 (2021). https://doi.org/10.1109/ACCESS.2021.3050934
6. Yanes Luis, S., Reina, D.G., Marin, S.L.T.: A deep reinforcement learning approach for the patrolling problem of water resources through autonomous surface vehicles: the Ypacarai lake case. IEEE Access **8**, 204076–204093 (2020). https://doi.org/10.1109/access.2020.3036938
7. Yanes Luis, S., Reina, D.G., Marin, S.L.T.: A multiagent deep reinforcement learning approach for path planning in autonomous surface vehicles: the Ypacaraí lake patrolling case. IEEE Access **9**, 17084–17099 (2021). https://doi.org/10.1109/access.2021.3053348
8. Arzamendia, M., Gregor, D., Reina, D.G., Toral, S.L.: An evolutionary approach to constrained path planning of an autonomous surface vehicle for maximizing the covered area of Ypacarai lake. Soft Comput. **23**(5), 1723–1734 (2019). https://doi.org/10.1007/s00500-017-2895-x
9. Arzamendia, M., Gutierrez, D., Toral, S., et al.: Intelligent online learning strategy for an autonomous surface vehicle in lake environments using evolutionary computation. IEEE Intell. Transp. Syst. Mag. **11**(4), 110–125 (2019). https://doi.org/10.1109/MITS.2019.2939109
10. Rückin, J., Jin, L., Popović, M.: Adaptive informative path planning using deep reinforcement learning for UAV-based active sensing, pp. 1–7 (2021). https://arxiv.org/abs/2109.13570
11. Rasmussen, C.E., Williams, C.K.I.: Gaussian Processes for Machine Learning, pp. 202–210. The MIT Press (2006). ISBN026218253X

Tuning ForestDisc Hyperparameters: A Sensitivity Analysis

Maissae Haddouchi$^{(\boxtimes)}$ 🆔 and Abdelaziz Berrado

AMIPS Team, Ecole Mohammadia d'Ingénieurs (EMI), Mohammed V University
in Rabat, Rabat, Morocco
maissaehaddouchi@research.emi.ac.ma, berrado@emi.ac.ma

Abstract. This paper presents and analyzes ForestDisc, a discretization method based on tree ensemble and moment matching optimization.
ForestDisc is a supervised and multivariate discretizer that transforms
continuous attributes into categorical ones following two steps. At first,
ForestDisc extracts for each continuous attribute the ensemble of split
points learned while constructing a Random Forest model. It then constructs a reduced set of split points based on moment matching optimization. Previous works showed that ForestDisc enables an excellent
performance compared to 22 popular discretizers. This work analyzes
ForestDisc performance sensitivity to its tunning parameters and provides some guidelines for users when using the ForestDisc package.

Keywords: Discretization · Optimization · Classification

1 Introduction

Discretization is a key pre-processing step in Machine Learning (ML). It is used
to reduce the complexity of the data space and to improve the performance and
efficiency of ML tasks [20]. Furthermore, it is a required pre-processing step for
several ML algorithms which only support categorical features. The usefulness
of discretization in ML, especially prior to classification tasks, has led to the
emergence of different discretization approaches. The literature has classified
them based on different dimensions, as being supervised versus unsupervised
and multivariate versus univariate [7,19,24,28].

Supervised discretization approaches consider the target attribute in the discretization process, which is expected to make them more "knowledgeable" in
determining the best splits than their "blind" unsupervised counterparts [1,24].
They use different metrics to minimize the loss of information between the discretized attributes and the target attribute on the one hand, and the number of
split points on the other hand [6,26]. Several comparative studies reported better performance for supervised discretization methods compared to unsupervised
ones [1,24].

Multivariate discretizers consider all data attributes simultaneously during
their processing, while univariate discretizers consider attributes one at a time.

B. Dorronsoro et al. (Eds.): OLA 2022, CCIS 1684, pp. 25–36, 2022.
https://doi.org/10.1007/978-3-031-22039-5_3

Multivariate discretization is sensitive to the data's correlation structure, unlike unsupervised one [22].

ForestDisc, a supervised and multivariate discretization proceeds in two main steps: First, it extracts the split points generated while learning a Random Forest (RF) model and then builds a set of cut points for each continuous attribute to produce its partition into bins. This second step relies on moment matching optimization to identify a set of cut points that optimally matches the statistical properties of the ensemble of split points generated through the RF model. A previous work [20] has demonstrated that ForestDisc reached an excellent performance compared to 20 other discretizers. The comparative analysis took into consideration different metrics. These metrics included the number of resulting bins per variable and the execution time needed for discretizing it. They also reported the performance of classifiers when discretization is applied before the classification task. This study was performed using 50 benchmark datasets and six well-known classifiers.

ForestDisc is available as an R package: **ForestDisc** [12, 13]. The first step in ForestDisc is processed using an RF model with 50 trees. The second step uses the first four moments in the moment matching approach and Nelder-Mead as an optimization algorithm.

In this work, we analyze the sensitivity of ForestDisc performance to the number of trees, the number of moments used, and the optimization algorithm used.

Accordingly, we present in the next section, ForestDisc related work. In Sect. 3, we present ForestDisc sensitivity analysis to its tunning parameters. Finally, we provide in the last section the conclusion.

2 Related Work

2.1 ForestDisc Algorithm

ForestDisc processes discretization in two stages. In the first stage (Algorithm 1), ForestDisc generates the ensemble of split points learned by an RF model. In the second stage (Algorithm 2), ForestDisc uses this ensemble and returns the set of cut points based on moment matching optimization.

Let **Data** represent a dataset. Let **AttCont** be the set of its continuous attributes. Let $S = \{S_A\}_{A \in AttCont}$ be the ensemble of splits values that would be learned throught an RF model. Each set S_A corresponds to an attribute A. Let $C = \{C_A\}_{A \in AttCont}$ be the ensemble of cut points that would be learned, where C_A is the set of cut points discretizing the attribute A. Let K_{max} be the maximum value allowed for C_A cardinality.

2.2 ForestDisc Tunning Parameters

We analyze in this section ForestDisc performance variability depending on the number of trees used in the RF model, the non-linear optimization algorithm

---

Algorithm 1. ForestDisc Algorithm - Step I

Input: *Data*
Output: *S*

Initialization: $S = null$

Fit **Random Forest** to **Data**
Return: $RF = \{T_1, ..., T_{n_{RF}}\}$ ▷ ensemmble of trees of cardinality n_{RF}
for each $A \in AttCont$ **do**
    Initialize: $S_A = null$
    for each *tree* **T** *in* **RF do**
        Extract S_A^T ▷ set of split values in tree T
        Update $S_A = S_A \cup S_A^T$
    end for
end for
Return: $S = \{S_A\}_{A \in AttCont}$

Algorithm 2. ForestDisc Algorithm - Step II

Input: *Data, AttCont, S, K_{max}*   ▷ K_{max} set by default to *10*
Output: *C, DataDisc*

Initialization: $C = null$

▷ We will solve, in the following, the moment matching problem (MMP) with n_m
moments
for each $A \in AttCont$ **do**
    for $j \in \{0...n_m\}$ **do**
        Compute S_A moment of order j: $m_A^j = \frac{\sum_{i=1}^{n_A} S_{Ai}^j}{n_A}$ ▷ n_A is S_A cardinality
        Compute S_A weight of order j: $w_A^j = \frac{1}{max(max(S_A), -min(S_A), 1)^{2j}}$
    end for
    for $k \in \{2...K_{max}\}$ **do**
        Solve the moment matching problem:
        Objective function $MMP(A,k)$: $min(P_k, X_k) \sum_{j=0}^{n_m} w_A^j \left(m_{X_k}^j - m_A^j\right)^2$
        where $m_{X_k}^j = \sum_{i=1}^{k} p_i x_i^j$, and $X_k = \{x_1, ..., x_k\}$ and $P_k = \{p_1, .., p_k\}$
        are the decision variables.

        Constraints: $\sum_{i=1}^{k} p_i - 1 = 0$ and $min(S_A) \leq x_i \leq max(S_A)$
        and $0 \leq p_i \leq 1$ for $i=1...k$

        Return: X_k^* and P_k^* the solution to $MMP(A,k)$, and Opt_k^* the optimum value
    end for
    Return: $X_A = \{X_k^*\}_{k \in \{2..K_{max}\}}$ and $Opt_A = \{Opt_k^*\}_{k \in \{2..K_{max}\}}$
    Select k_{opt} the value k corresponding to the minimum value in the set Opt_A
    Return: $C_A = X_{k_{opt}}^*$
end for
Return: $C = \{C_A\}_{A \in ContAttr}$

---

used for solving the MMP (Algorithm 2), and the number of moments considered in MMP. We consider in this work three tunning parameters for ForestDisc.

Number of Trees: ForestDisc is based on the decision trees grown by an RF model to build an ensemble of split points partitioning the continuous attributes in a supervised and multivariate way (Algorithm 1). RF is known, in the literature, as an efficient ensemble learning, robust against overfitting, and user-friendly [21]. However, the number of trees is an important tuning parameter that generally influences the performance of an RF model. The optimal number of trees is problem dependant and users generally resort to comparative analysis to assess how the number of trees impacts the RF performance. We will analyze in the following sections how ForestDisc performance varies depending on the number of trees used.

Optimization Algorithm: The moment matching problem introduced in Algorithm 2 is a non-linear optimization (NLO) problem. NLO [2] solves optimization problems where the objective function or some of the equality or inequality constraints are non-linear. We compared in previous work [14], the performance of 22 NLO Algorithm on solving the discretization based on MMP. The discretizer used in [14] is a simplified version of ForestDisc. It is a univariate and unsupervised discretization method, mapping each continuous attribute to a categorical attribute based on MMP. The performance of the 22 NLO Algorithms was compared based on multiple measures. The empirical results showed that the Nelder-Mead Simplex Algorithm [23] (Neldermead) achieved the best tradeoff between intrinsic and extrinsic measures [20]. The DIviding RECTangles (locally biased) algorithm [16] (DIRECTL) realized the second-best tradeoff, and the Sequential Least-Squares Quadratic Programming algorithm [17,18] (SLSQP) performed the best matching (optimum value). We compare in the following sections how ForestDisc performance changes regarding these 3 NLO algorithms.

Number of Moments. The moment matching mapping used in ForestDisc is based on the approach proposed in [46]. Authors in this work have conducted moment matching based on the first four moments to generate a limited number of discrete outcomes. Furthermore, the first four moments are widely known to characterize important statistics of random variables, namely, the mean, the variance, skewness, and kurtosis, which convey practical insights about distributions of random variables. ForestDisc proposed in [20] also uses the four first moments. This being said, it is worthwhile to investigate to what extent changing the number of moments in MMP (Algorithm 2) would impact the performance of the resulting discretization. The following sections investigate the sensitivity of ForestDisc to the number of first moments used in MMP.

3   ForestDisc Sensitivity Analysis

3.1   Experimental Set up

In this section, we evaluate the sensitivity of ForestDisc to its tunning parameters: the number of trees, the number of moments, and the optimization algorithm used. We use the following metrics in this comparative analysis.

The first one is the optimum value, which is the solution to the moment matching problem described in Algorithm 2. The second one is the execution time required for discretizing an attribute. The third one is the number of resulting bins per discretized attribute.

The fourth and fifth metrics concern the predictive performance of classifiers pre-processed by discretization. The predictive performance is assessed via accuracy and F1 measures (details about the computation of each of these metrics can be found in [20]). Five well known classifiers are used in this comparative study: Classification And Regression Trees (CART) [4], Random Forest (RF) [3], Tree Boosting (Boosting) [5,9], Optimal weighted K-nearest neighbor classifier (KNNC) [25], and Naive Bayes Classifier (NaiveBayes) [10]. Their respective R functions/packages are: rpart/rpart, randomForest/randomForest, xgboost/xgboost, kknn/kknn, and naiveBayes/e1071. The default parameters were used for each of the aforementioned functions. RF and Boosting were performed using 200 trees.

We also use the Wilcoxon signed-rank test [27] to statistically compare the different results. We adopt the approach used in [11,20] to perform pairwise comparisons of the metrics considered in this study. The results are summarized by counting the times each method outperforms, ties, and underperforms.

We analyze the different metrics by using benchmark datasets taken from the UCI Machine Learning [8] and Keel data sets repository [15]. Each data set is processed 10 times using Monte Carlo cross-validation procedure (refer to [20] for more details).

3.2   ForestDisc Sensitivity to the Number of Trees
and the Non-linear Optimization Algorithm Used

We report in Fig. 1 the average results on accuracy and F1 score over the five classifiers, the 50 datasets used in [20], and the 10 iterations. These results show that the Neldermed algorithm outperforms the DIRECTL and SLSQP algorithms regardless of the number of trees used. Moreover, increasing the number of trees does not seem to induce an improvement in predictive performance regardless of the NLO algorithm used. The Wilcoxon test results reported in Table 1 confirm this conclusion.

Figure 2 displays the average results on the execution time per discretized variable and the number of intervals per variable. Based on these results, the Neldemead algorithm outperforms the other NLO algorithms in terms of execution time regardless of the number of trees used. In addition, the execution time increases as the number of trees increases regardless of the NLO used.

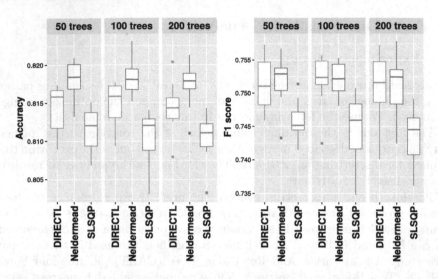

Fig. 1. Variation of the average accuracy and F1 score on the testing sets depending on the number of trees and the NLO algorithms used

Table 1. Wilcoxon signed-rank test scoring in global accuracy and F1 score depending on the number of trees and the NLO algorithms used (on the testing sets)

Accuracy				F1 measure			
Method	Wins	Ties	Losses	Method	Wins	Ties	Losses
Neldermead50	7	1	0	Neldermead50	7	1	0
Neldermead100	7	1	0	Neldermead100	7	1	0
Neldermead200	6	0	2	Neldermead200	5	1	2
DIRECTL50	3	2	3	DIRECTL100	3	3	2
DIRECTL100	3	2	3	DIRECTL50	3	2	3
DIRECTL200	3	2	3	DIRECTL200	3	2	3
SLSQP50	2	0	6	SLSQP50	1	1	6
SLSQP100	0	1	7	SLSQP100	0	2	6
SLSQP200	0	1	7	SLSQP200	0	1	7

The number of intervals does not seem to vary according to the number of trees. Neldermead algorithm tends to produce the highest number of bins (between 6.5 and 7 intervals), followed by DIRECTL (between 4 and 5), and lastly by SLSQP (slightly less than 4). The Wilcoxon results reported in Table 2 are in accordance with these conclusions.

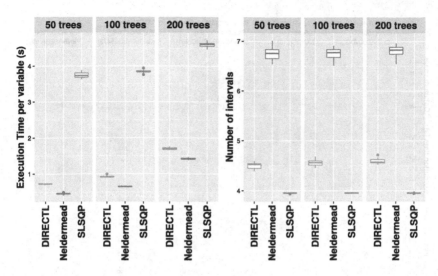

Fig. 2. Variation of the average execution time and number of intervals depending on the number of trees and the NLO algorithms used

Table 2. Wilcoxon signed-rank test scoring in Execution time and number of intervals depending on the number of trees and the NLO algorithms used (on the testing sets)

Execution time				Number of intervals			
Method	Wins	Ties	Losses	Method	Wins	Ties	Losses
Neldermead50	8	0	0	SLSQP50	6	2	0
Neldermead100	7	0	1	SLSQP100	6	2	0
DIRECT50	6	0	2	SLSQP200	6	2	0
DIRECT100	5	0	3	DIRECT50	5	0	3
Neldermead200	4	0	4	DIRECT100	4	0	4
DIRECT200	3	0	5	DIRECT200	3	0	5
SLSQP50	2	0	6	Neldermead50	1	1	6
SLSQP100	1	0	7	Neldermead100	1	1	6
SLSQP200	0	0	8	Neldermead200	0	0	8

3.3   ForestDisc Sensitivity to the Number of Moments Used

In this section, we analyze the sensitivity of ForestDisc to the number of first moments used in the MMP. This analysis is performed by varying the number of moments from 2 to 7, on a selection of 15 benchmark data sets used in a previous work [13].

Figure 3 reports the average results on the optimum values (MMP solutions), the execution time per discretized variable, and the number of intervals per variable. These results show that the optimum value is the greatest when 4 moments are used. Nevertheless, the use of 4 moments allows for a very good

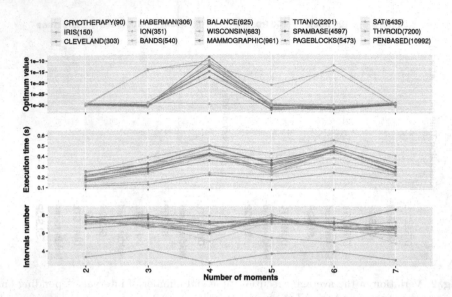

Fig. 3. Discretization performance variation depending on the number of moments (optimum value, execution time, and number of bins)

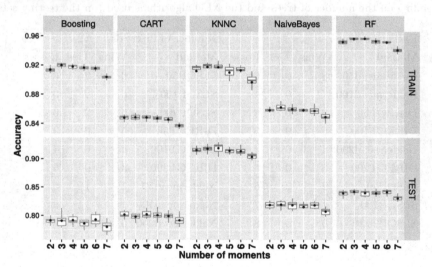

Fig. 4. Classifiers accuracy variation depending on the number of moments (on the training and testing sets). Bold dots show the mean values.

moment matching since the value of the optimum value remains very small (less than 1E−10 in general). The execution time tends to be slightly higher when 4 or 6 moments are used. Finally, the number of intervals tends to be slightly smaller when 4 or 6 moments are used.

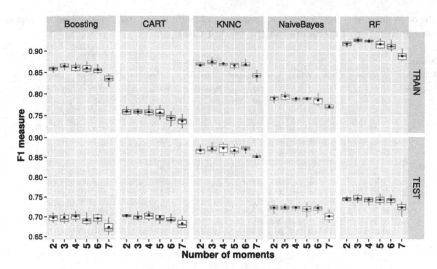

Fig. 5. Classifiers F1 score variation depending on the number of moments (on the training and testing sets). Bold dots show the mean values.

Figures 4 and 5 display the variation of 5 classifiers' accuracy and F1 score depending on the number of moments used. The results are plotted on the training and testing sets. Table 3 reports the Wilcoxon signed test scores computed using accuracy and F1 scores on the testing sets. The results show that there is not a significant difference in predictive performance depending on the number of moments used, except for 7 moments (the worst results for all the classifiers). Using 4 moments seems to be a good alternative for all the classifiers (see Table 3). However, we think that we should use a more extensive analysis to have a robust conclusion about the impact of the number of moments used on the classifiers' predictive performance.

Table 3. Wilcoxon signed-rank test scoring in classifiers accuracy and F1 score depending on the number of moments (on the testing sets)

Accuracy					F1 measure				
Classifier	Moments Nbr	Wins	Ties	Losses	Classifier	Moments Nbr	Wins	Ties	Losses
Boosting	4	2	3	0	Boosting	4	2	3	0
	6	2	3	0		3	1	4	0
	3	1	4	0		6	1	4	0
	2	1	4	0		2	1	4	0
	5	1	2	2		5	1	3	1
	7	0	0	5		7	0	0	5
CART	4	2	3	0	CART	4	3	2	0
	6	1	4	0		2	2	3	0
	5	1	4	0		5	1	4	0
	2	1	4	0		3	1	3	1
	3	1	3	1		6	1	2	2
	7	0	0	5		7	0	0	5
KNNC	4	3	2	0	KNNC	4	1	4	0
	3	3	2	0		3	1	4	0
	2	1	4	0		2	1	4	0
	5	1	2	2		5	1	4	0
	6	1	2	2		6	1	4	0
	7	0	0	5		7	0	0	5
NaiveBayes	3	2	3	0	NaiveBayes	4	1	4	0
	4	1	4	0		3	1	4	0
	6	1	4	0		2	1	4	0
	2	1	4	0		5	1	4	0
	5	1	3	1		6	1	4	0
	7	0	0	5		7	0	0	5
RF	3	2	3	0	RF	3	2	3	0
	4	1	4	0		4	1	4	0
	6	1	4	0		6	1	4	0
	2	1	4	0		2	1	4	0
	5	1	3	1		5	1	3	1
	7	0	0	5		7	0	0	5

4   Conclusion

In this work, we have investigated the sensitivity of the ForestDisc discretizer to its tunning parameters. ForestDisc discretizes continuous attributes in two steps. First, it uses the ensemble of decision trees grown by an RF model to build an ensemble of split points partitioning the continuous attributes. It then relies on moment matching optimization to return a reduced set of cut points for discretizing each continuous attribute. A previous work [20] has demonstrated that ForestDisc reached an excellent performance compared to 20 other discretizers, based on extensive analysis on 50 benchmark datasets and six well-known classifiers.

We have analyzed, in this work, the sensitivity of ForestDisc performance to its tunning parameters by varying the number of trees from 50 to 200 and the number of moments from 2 to 7. We have also used two other alternatives for the optimization algorithm used. This preliminary analysis has shown that using 50 trees, four moments, and the Neldermead algorithm in the ForestDisc framework leads to the best performance. We think, however, that we should, in future work, expand the ranges of the number of trees and the number of moments to make a robust conclusion about the sensitivity of ForestDisc to these two parameters.

References

1. Agre, G.: On supervised and unsupervised discretization. Cybern. Inf. Technol. (2002)
2. Bazaraa, M.S., Sherali, H.D., Shetty, C.M.: Nonlinear Programming: Theory and Algorithms, 3rd edn. Wiley-Interscience, Hoboken (2006). oCLC: ocm61478842
3. Breiman, L.: Random forests. Mach. Learn. **45**(1), 5–32 (2001). https://doi.org/10.1023/A:1010933404324
4. Breiman, L., Friedman, J.H., Olshen, R.A., Stone, C.J.: Classification and Regression Trees. Wadsworth & Brooks/Cole Advanced Books & Software, Monterey (1984). 358 p., the wadsworth statistics/probability series edn. (1884)
5. Chen, T., Guestrin, C.: XGBoost: a scalable tree boosting system. In: Proceedings of the 22nd ACM SIGKDD International Conference on Knowledge Discovery and Data Mining - KDD 2016, pp. 785–794. ACM Press, San Francisco (2016). https://doi.org/10.1145/2939672.2939785
6. Ching, J., Wong, A., Chan, K.: Class-dependent discretization for inductive learning from continuous and mixed-mode data. IEEE Trans. Pattern Anal. Mach. Intell. **17**(7), 641–651 (1995). https://doi.org/10.1109/34.391407
7. Dougherty, J., Kohavi, R., Sahami, M.: Supervised and unsupervised discretization of continuous features. In: Machine Learning Proceedings 1995, pp. 194–202. Elsevier (1995). https://doi.org/10.1016/B978-1-55860-377-6.50032-3
8. Dua, D., Graff, C.: UCI machine learning repository (2017)
9. Friedman, J.H.: Greedy function approximation: a gradient boosting machine. Ann. Stat. **29**, 1189–1232 (2000)
10. Friedman, N., Geiger, D., Goldszmidt, M.: Bayesian network classifiers. Mach. Learn. **29**(2), 131–163 (1997). https://doi.org/10.1023/A:1007465528199
11. Garcia, S., Luengo, J., Sáez, J.A., López, V., Herrera, F.: A survey of discretization techniques: taxonomy and empirical analysis in supervised learning. IEEE Trans. Knowl. Data Eng. **25**(4), 734–750 (2013). https://doi.org/10.1109/TKDE.2012.35
12. Haddouchi, M.: ForestDisc: forest discretization. R package version 0.1.0 (2020). https://CRAN.R-project.org/package=ForestDisc
13. Haddouchi, M., Berrado, A.: An implementation of a multivariate discretization for supervised learning using Forestdisc, pp. 1–6 (2020). https://doi.org/10.1145/3419604.3419772
14. Haddouchi, M., Berrado, A.: Discretizing continuous attributes for machine learning using nonlinear programming. Int. J. Comput. Sci. Appl. **18**(1), 26–44, 20 (2021)
15. Alcalá-Fdez, J., et al.: KEEL data-mining software tool: data set repository, integration of algorithms and experimental analysis framework. J. Multiple-Valued Log. Soft Comput. **17**(2–3), 255–287 (2011)
16. Jones, D.R., Perttunen, C.D., Stuckman, B.E.: Lipschitzian optimization without the Lipschitz constant. J. Optim. Theory Appl. **79**(1), 157–181 (1993). https://doi.org/10.1007/BF00941892

17. Kraft, D.: A Software Package for Sequential Quadratic Programming. Deutsche Forschungs- Und Versuchsanstalt Für Luft- Und Raumfahrt Köln: Forschungsbericht, Wiss. Berichtswesen d. DFVLR (1988)
18. Kraft, D., Munchen, I.: Algorithm 733: TOMP - Fortran modules for optimal control calculations. ACM Trans. Math. Soft **20**, 262–281 (1994)
19. Liu, H., Hussain, F., Tan, C.L., Dash, M.: Discretization: an enabling technique. Data Min. Knowl. Disc. **6**, 393–423 (2002)
20. Maissae, H., Abdelaziz, B.: A novel approach for discretizing continuous attributes based on tree ensemble and moment matching optimization. Int. J. Data Sci. Anal. (2022). https://doi.org/10.1007/s41060-022-00316-1
21. Haddouchi, M., errado, A.: A survey of methods and tools used for interpreting random forest, pp. 1–6 (2019). https://doi.org/10.1109/ICSSD47982.2019.9002770
22. Mehta, S., Parthasarathy, S., Yang, H.: Toward unsupervised correlation preserving discretization. IEEE Trans. Knowl. Data Eng. **17**(9), 1174–1185 (2005). https://doi.org/10.1109/TKDE.2005.153
23. Nelder, J.A., Mead, R.: A simplex method for function minimization. Comput. J. **7**, 308–313 (1965). https://doi.org/10.1093/comjnl/7.4.308
24. Ramırez-Gallego, S., Garcıa, S., Martınez-Rego, D., Benıtez, J.M., Herrera, F.: Data discretization: taxonomy and big data challenge, p. 26 (2016)
25. Samworth, R.J.: Optimal weighted nearest neighbour classifiers. Ann. Stat. **40**(5), 2733–2763 (2012). https://doi.org/10.1214/12-AOS1049
26. Wang, C., Wang, M., She, Z., Cao, L.: CD: a coupled discretization algorithm. In: Tan, P.-N., Chawla, S., Ho, C.K., Bailey, J. (eds.) PAKDD 2012. LNCS (LNAI), vol. 7302, pp. 407–418. Springer, Heidelberg (2012). https://doi.org/10.1007/978-3-642-30220-6_34
27. Wilcoxon, F.: Individual comparisons by ranking methods. Biometr. Bull. **1**(6), 80 (1945). https://doi.org/10.2307/3001968
28. Yang, Y., Webb, G.I., Wu, X.: Discretization methods. In: Maimon, O., Rokach, L. (eds.) Data Mining and Knowledge Discovery Handbook, pp. 101–116. Springer, Boston (2010). https://doi.org/10.1007/978-0-387-09823-4_6

Multi-objective Hyperparameter Optimization with Performance Uncertainty

Alejandro Morales-Hernández[1,2,3](✉) [iD], Inneke Van Nieuwenhuyse[1,2,3] [iD], and Gonzalo Nápoles[4] [iD]

[1] Core Lab VCCM, Flanders Make, Limburg, Belgium
[2] Research Group Logistics, Hasselt University, Agoralaan Gebouw D, Diepenbeek, 3590 Limburg, Belgium
[3] Data Science Institute, Hasselt University, Agoralaan Gebouw D, Diepenbeek, 3590 Limburg, Belgium
{alejandro.moraleshernandez,inneke.vannieuwenhuyse}@uhasselt.be
[4] Department of Cognitive Science and Artificial Intelligence, Tilburg University, Tilburg, The Netherlands
g.r.napoles@uvt.nl

Abstract. The performance of any Machine Learning algorithm is impacted by the choice of its hyperparameters. As training and evaluating a ML algorithm is usually expensive, the hyperparameter optimization (HPO) method needs to be computationally efficient to be useful in practice. Most of the existing approaches on multi-objective HPO use evolutionary strategies and metamodel-based optimization. However, few methods account for uncertainty in the performance measurements. This paper presents results on multi-objective HPO with uncertainty on the performance evaluations of the ML algorithms. We combine the sampling strategy of Tree-structured Parzen Estimators (TPE) with the metamodel obtained after training a Gaussian Process Regression (GPR) with heterogeneous noise. Experimental results on three analytical test functions and three ML problems show the improvement in the hypervolume obtained, when compared with HPO using stand-alone multi-objective TPE and GPR.

Keywords: Hyperparameter optimization · Multi-objective optimization · Bayesian optimization · Uncertainty

1 Introduction

In Machine Learning (ML), an hyperparameter is a parameter that needs to be specified before training the algorithm: it influences the learning process, but it is not optimized as part of the training algorithm. The time needed to train a ML algorithm with a given hyperparameter configuration on a given dataset may already be substantial, particularly for moderate to large datasets,

B. Dorronsoro et al. (Eds.): OLA 2022, CCIS 1684, pp. 37–46, 2022.
https://doi.org/10.1007/978-3-031-22039-5_4

so the HPO algorithm should be as efficient as possible in detecting the optimal hyperparameter setting.

Many of the current algorithms in the literature focus on optimizing a single (often error-based) objective [2,10,13]. In practical applications, however, it is often required to consider the trade-off between two or more objectives, such as the error-based performance of a model and its resource consumption [7], or objectives relating to different types of error-based performance measures [5]. The goal in multi-objective HPO is to obtain the *Pareto-optimal* solutions, i.e., those hyperparameter values for which none of the performance measures can be improved without negatively affecting any other.

In the literature, most HPO approaches take a deterministic perspective using the mean value of the performance observed in subsets of data (cross validation protocol). However, depending on the chosen sets, the outcome may differ: a single HP configuration may thus yield different results for each performance objective, implying that the objective contains uncertainty (hereafter referred to as *noise*). We conjecture that a HPO approach that considers this uncertainty will outperform alternative approaches that assume the relationships to be deterministic. Stochastic algorithms (such as [3,4]) can potentially be useful for problems with heterogeneous noise (the noise level varies from one setting to another). To the best of our knowledge, such approaches have not yet been studied in the context of HPO optimization. The main contributions of our approach include:

- Multi-objective optimization using a Gaussian Process Regression (GPR) surrogate that explicitly accounts for the heterogeneous noise observed in the performance of the ML algorithm.
- The selection of infill points according to the sampling strategy of multi-objective TPE (MOTPE), and the maximization of an infill criterion. This method allows sequential selection of hyperparameter configurations that are likely to be non-dominated, and that yield the largest expected improvement in the Pareto front.

The remainder of this article is organized as follows. Section 2 discusses the basics of GPR and MOTPE. Section 3 presents the algorithm. Section 4 describes the experimental setting designed to evaluate the proposed algorithm, and Sect. 5 shows the results. Finally, Sect. 6 summarizes the findings and highlights some future research directions.

2   GPR and TPE: Basics

Gaussian Process Regression (GPR) (also referred to as *kriging*, [14]) is commonly used to model an unknown target function. The function value prediction at an unsampled point $\mathbf{x}^{(*)}$ is obtained through the conditional probability $P(f(\mathbf{x}^{(*)})|\mathbf{X},\mathbf{Y})$ that represents how likely the response $f(\mathbf{x}^{(*)})$ is, given that we observed the target function at n input locations $\mathbf{x}^{(i)}, i = 1, \ldots, n$ (contained in matrix $\mathbf{X}$), yielding function values $\mathbf{y}^{(i)}, i = 1, \ldots, n$ (contained in matrix $\mathbf{Y}$)

that may or may not be affected by noise. Ankenman et al. [1] provides a GPR model (referred to as *stochastic kriging*) that takes into account the heterogeneous noise observed in the data, and models the observed response value in the r-th replication at design point $\mathbf{x}^{(i)}$ as:

$$f_r(\mathbf{x}^{(i)}) = m(\mathbf{x}^{(i)}) + M(\mathbf{x}^{(i)}) + \epsilon_r(\mathbf{x}^{(i)}) \tag{1}$$

where $m(\mathbf{x})$ represents the mean of the process, $M(\mathbf{x})$ is a realization of a Gaussian random field with mean zero (also referred to as the *extrinsic uncertainty* [1]), and $\epsilon_r(\mathbf{x}^{(i)})$ is the *intrinsic uncertainty* observed in replication r. Popular choices for $m(\mathbf{x})$ are $m(\mathbf{x}) = \sum_h \beta_h f_h(\mathbf{x})$ (where the $f_h(\mathbf{x})$ are known linear or nonlinear functions of $\mathbf{x}$, and the β_h are unknown coefficients to be estimated), $m(\mathbf{x}) = \beta_0$ (an unknown constant to be estimated), or $m(\mathbf{x}) = 0$. $M(\mathbf{x})$ can be seen as a function, randomly sampled from a space of functions that, by assumption, exhibit spatial correlation according to a covariance function (also referred to as *kernel*).

Whereas GPR models the probability distribution of $f(\mathbf{x})$ given a set of observed points ($P(f(\mathbf{x})|\mathbf{X}, \mathbf{Y})$), TPE tries to model the probability of sampling a point that is directly associated to the set of observed responses ($P(\mathbf{x}|\mathbf{X}, \mathbf{Y})$) [2]. TPE defines $P(\mathbf{x}|\mathbf{X}, \mathbf{Y})$ using two densities:

$$P(\mathbf{x}|\mathbf{X}, \mathbf{Y}) = \begin{cases} l(x) & if \quad f(x) < y^*, \mathbf{x} \in \mathbf{X} \\ g(x) & o.w \end{cases} \tag{2}$$

where $l(x)$ is the density estimated using the points $\mathbf{x}^{(i)}$ for which $f(\mathbf{x}^{(i)}) < y^*$, and $g(x)$ is the density estimated using the remaining points. The value y^* is a user-defined quantile γ (splitting parameter of Algorithm 1 in [11]) of the observed $f(\mathbf{x})$ values, so that $P(f(\mathbf{x}) < y^*) = \gamma$. Here, we can see l as the density of the hyperparameter configurations that may have the best response. A multi-objective implementation of TPE (MOTPE) was proposed by [11]; this multi-objective version splits the known observations according to their nondomination rank. Contrary to GPR, neither TPE nor MOTPE provide an estimator of the response at unobserved hyperparameter configurations.

3 Proposed Algorithm

The algorithm (Fig. 1) starts by evaluating an initial set of hyperparameter vectors through a Latin hypercube sample; simulation replications are used to estimate the objective values at these points. We then perform two processes in parallel. On the one hand, we use the augmented Tchebycheff scalarization function [9] (with a random combination of weights) to transform the multiple objectives into a single objective using these training data. Throughout this article, we will assume that the individual objectives need to be minimized; hence, the resulting scalarized objective function also needs to be minimized. We then train a (single) stochastic GPR metamodel on these scalarized objective function outcomes; the replication outcomes are used to compute the variance of this scalarized objective tive.

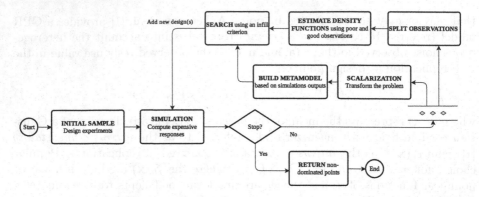

Fig. 1. Proposed multi-objective HPO using GPR with heterogeneous noise and TPE to sample the search space

At the same time, we perform the splitting process used by [11] to divide the hyperparameter vectors into two subsets (those yielding "good" and "poor" observations) to estimate the densities $l(x)$ and $g(x)$ for each separate input dimension (Eq. 2). To that end, our approach uses a greedy selection according to the nondomination rank of the observations, and controlled by the parameter γ[1]. The strategy thus preferably selects the HP configurations with highest nondomination rank to enter in the "good" subset.

Using the densities $l(x)$, we randomly select a candidate set of n_c configurations for each input dimension. These individual values are sorted according to their log-likelihood ratio $log\frac{l(x)}{g(x)}$, such that the higher this score, the larger the probability that the input value is sampled under $l(\mathbf{x}_i)$ (and/or the lower the probability under $g(\mathbf{x}_i)$). Instead of selecting the single configuration with highest score on each dimension (as in [2,11]), we compute the aggregated score $AS(\mathbf{x}) = \sum_{i=1}^{d} log\frac{l(x_i)}{g(x_i)}$ for each configuration, and select the one that maximizes the *Modified Expected Improvement* (MEI) [12] of the scalarized objective function in the set of configurations Q with an aggregated score greater than zero (see Eq. 3).

$$\arg\max_{\mathbf{q} \in Q} (\widehat{Z}_{\min} - \widehat{Z}_{\mathbf{q}})\Phi(\frac{\widehat{Z}_{\min} - \widehat{Z}_{\mathbf{q}}}{\widehat{s}_{\mathbf{q}}}) + \widehat{s}_{\mathbf{q}}\phi(\frac{\widehat{Z}_{\min} - \widehat{Z}_{\mathbf{q}}}{\widehat{s}_{\mathbf{q}}}) \, , Q = \{\mathbf{x} \,|\, AS(\mathbf{x}) > 0\}$$

(3)

where $\widehat{Z}_{\min}$ is the stochastic kriging prediction at $\mathbf{x}_{\min}$ (i.e. the hyperparameter configuration with the lowest sample mean among the already known configurations), $\phi(\cdot)$ and $\Phi(\cdot)$ are the standard normal density and standard normal distribution function respectively, the $\widehat{Z}_{\mathbf{q}}$ is the stochastic kriging prediction at configuration $\mathbf{q}$, and $\widehat{s}_{\mathbf{q}}$ is the ordinary kriging standard deviation for that configuration [15]. The search using MEI focuses on new points located in promising

---
[1] Notice that both in [11] and in our algorithm, the parameter γ represents a percentage of the known observations that may be considered as "good".

regions (i.e., with low predicted responses; recall that we assume that the scalarized objective need to be minimized), or in regions with high metamodel uncertainty (i.e., where little is known yet about the objective function). Consequently, the sampling behavior automatically trades off exploration and exploitation of the configuration search space.

Once a new hyperparameter configuration has been selected as infill point, the ML algorithm is trained on this configuration, yielding (again) noisy estimates of the performance measures. Following this infill strategy, we choose that configuration for which we expect the biggest improvement in the scalarized objective function, among the configurations that are likely to be non-dominated.

4   Numerical Simulations

In this section, we evaluate the performance of the proposed algorithm for solving multi-objective optimization problems (GP_MOTPE), comparing the results with those that would be obtained by using GP modelling and MOTPE individually. In a first experiment, we analyze the performance on three well-known bi-objective problems (ZDT1, WFG4 and DTLZ7 with input dimension $d = 5$; see [6]), to which we add artificial heterogeneous noise (as in [4]). More specifically, we obtain noisy observations $\widetilde{f}_p^j(\mathbf{X}_i) = f_j(\mathbf{X}_i) + \epsilon_p(\mathbf{X}_i), p = \{1, \ldots, r\}, j = \{1, \ldots, m\}$, with $\epsilon_p(\mathbf{X}_i) \sim \mathcal{N}(0, \tau_j(\mathbf{X}_i))$. The standard deviation of the noise $(\tau_j(\mathbf{X}))$ varies for each objective between $0.01 \times \Omega^j$ and $0.5 \times \Omega^j$, where Ω^j is the range of objective j. In between these limits, $\tau_j(\mathbf{X})$ decreases linearly with the objective value: $\tau_j(\mathbf{X}) = a_j(f_j(\mathbf{X}) + b_j), \forall j \in \{1, \ldots, m\}$, where a and b are the linear coefficients obtained from the noise range [8].

Table 1. Details of the ML datasets

Dataset	ID	Inst. (Feat.)	Dataset	ID	Inst. (Feat.)
Balance-scale	997	625 (4)	Delta_ailerons	803	7129 (5)
Optdigits	980	5620 (64)	Heart-statlog	53	270 (13)
Stock	841	950 (9)	Chscase_vine2	814	468 (2)
Pollen	871	6848 (5)	Ilpd	41945	583 (10)
Sylvine	41146	5124 (20)	Bodyfat	778	252 (14)
Wind	847	6574 (14)	Strikes	770	625 (6)

In a second experiment, we test the algorithm on a number of OpenML datasets, shown in Table 1. We optimize five hyperparameters for a simple (one hidden layer) Multi-Layer Perceptron (MLP), two for a support vector machine (SVM), and five for a Decision Tree (DT) (see Appendix A). In each experiment, the goal is to find the HPO configurations that minimize classification error while simultaneously maximizing recall. In all experiments, we used 20% of the initial dataset as test set, and the remainder for HPO. We apply stratified *k-fold cross-validation* ($k = 10$) to evaluate each hyperparameter configuration.

We used a fixed, small number of iterations (100) as a stopping criterion in all algorithms; this keeps optimization time low, and resembles real-world optimization settings where limited resources (e.g., time) may exist. Table 2 summarizes the rest of the parameters used in the experiments.

Table 2. Summary of the parameters for the experiments

Setting	Problem	GP	MOTPE	GP_MOTPE
Initial design	Analytical fcts	LHS: $11d - 1$		
	HPO	Random sampling: $11d - 1$		
Replications	Analytical fcts	50		
	HPO	10		
Acquisition function		MEI	EI$_{\text{TPE}}$	MEI
Acquisition function optimization		**PSO***	Maximization on a candidate set	
Number of candidates to sample		–	$n_c = 1000, \gamma = 0.3$	
Kernel		Gaussian	–	Gaussian

*PSO algorithm (Pyswarm library): swarm size = 300, max iterations = 1800, cognitive parameter = 0.5, social parameter = 0.3, and inertia = 0.9

5   Results

Figure 3 shows the evolution of the hypervolume indicator during the optimization of the analytical test functions. The combined algorithm GP_MOTPE yields a big improvement over both GP and MOTPE algorithms for the ZDT1 and DTLZ7 functions, reaching a superior hypervolume already after a small number of iterations. Results also show that for ZDT1 and DTLZ7, the standard deviation on the final hypervolume obtained by GP and GP_MOTPE is small, which indicates that a Pareto front of similar quality is obtained regardless of the initial design. MOTPE, by contrast, shows higher uncertainty in the hypervolume results at the end of the optimization. For the concave Pareto front of WFG4, MOTPE provides the best results, while GP_MOTPE still outperforms GP (Fig. 2).

Table 3 shows the average rank of the optimization algorithms according to the hypervolume indicator. The experiments did not highlight significant differences between GP_MOTPE, GP and MOTPE ($p\_value = 0.565 > 0.05$ for the non-parametric Friedman test where H_0 states that the mean hypervolume of the solutions is equal). However, GP_MOTPE has the lowest average rank in the validation set, indicating that on average, the Pareto front obtained with our algorithm tends to outperform those found by GP and MOTPE individually, yielding a larger hypervolume.

Once the Pareto-optimal set of HP configurations has been obtained on the validation set, the ML algorithm (trained with those configurations) is evaluated on the test set. The difference between the hypervolume values obtained from

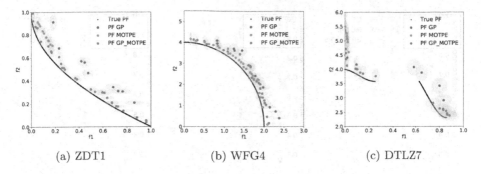

(a) ZDT1   (b) WFG4   (c) DTLZ7

Fig. 2. Observed Pareto front (PF) obtained at the end of a single macroreplication, for the analytical test functions. The uncertainty of each solution is shown by a shaded ellipse, and reflects the *mean ± std* of the simulation replications.

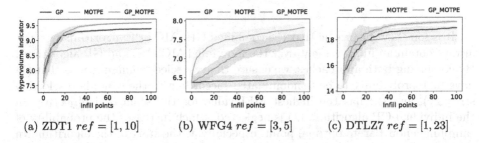

(a) ZDT1 *ref* = [1, 10]   (b) WFG4 *ref* = [3, 5]   (c) DTLZ7 *ref* = [1, 23]

Fig. 3. Hypervolume evolution during the optimization of the analytical test functions. Shaded area represents *mean ± std* of 13 macro-replications. Captions contain the reference point used to compute the hypervolume indicator

the validation and test set can be used as a measure of reliability: in general, one would prefer HP configurations that generate a similar hypervolume in the test set. Figure 4 shows that the difference between both hypervolume values is almost zero when GP_MOTPE is used, for all ML algorithms. In general, MOTPE and GP_MOTPE have the smallest (almost identical) mean absolute hypervolume difference (0.0444 and 0.0445 respectively), compared with that of GP (0.051). However, GP_MOTPE has the smallest standard deviation (0.054), followed by MOTPE (0.066) and GP (0.067).

Table 3. Average rank (given by the hypervolume indicator) of each algorithm

	Validation set			Test set		
	GP	MOTPE	GP_MOTPE	GP	MOTPE	GP_MOTPE
Avg. rank	2.125	1.9861	1.8889	2.1528	1.875	1.9722

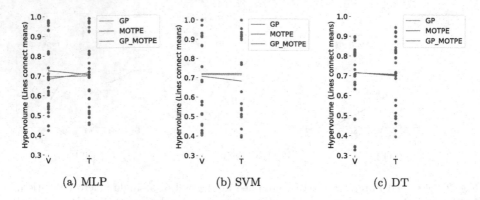

Fig. 4. Hypervolume generated by the HP configurations found using the validations set (V) and then evaluated with the test set (T)

It is somehow surprising that the combined GP_MOTPE algorithm does not always obtain an improvement over the individual MOTPE and GP algorithms. By combining both approaches, we ensure that we select configurations that (1) have high probability to be nondominated (according to the candidate selection strategy), and (2) has the highest MEI value for the scalarized objective. In the individual GP algorithm, (1) is neglected, which increases the probability of sampling a non-Pareto optimal point, especially at the start of the algorithm. In the original MOTPE algorithm, (2) is neglected, which may cause the algorithm to focus too much on exploitation, which increases the probability of ending up in a local optimum. We suspect that the MOTPE approach for selecting candidate points may actually be too restrictive: it will favor candidate points close to already sampled locations, inherently limiting the exploration opportunities the algorithm still has when optimizing MEI.

6   Concluding Remarks

In this paper, we proposed a new algorithm (GP_MOTPE) for multi-objective HPO of ML algorithms. This algorithm combines the predictor information (both predictor and predictor variance) obtained from a GPR model with heterogenous noise, and the sampling strategy performed by Multi-objective Tree-structured Parzen Estimators (MOTPE). In this way, the algorithm should select new points that are likely to be non-dominated, and that are expected to cause the maximum improvement in the scalarized objective function.

The experiments conducted report that our approach performed relatively well for the analytical test functions of study. It appears to outperform the pure GP algorithm in all analytical instances; yet, it does not always outperform the original MOTPE algorithm. Further research will focus on why this is the case, which may yield further improvements in the algorithm. In the HPO experiments, GP_MOTPE shows the best average rank w.r.t. the hypervolume computed on the validation set. In addition, it showed promising reliability properties (small changes in hypervolume when the ML algorithm is evaluated on the test set). Based upon these first results, we believe that the combination of GP and TPE is promising enough to warrant further research. The observation that it outperforms the pure GP algorithm (which used PSO to maximize the infill criterion) is useful in its own right, as the optimization of infill criteria is known to be challenging. Using MOTPE, a candidate set can be generated that can be evaluated efficiently, and which (from these first results) appears to yield superior results.

Acknowledgements. This research was supported by the Flanders Artificial Intelligence Research Program (FLAIR).

Appendix A. Setup of Hyperparameters in the HPO Experiments

HP	Description	Type	Range
Multilayer Perceptron (MLP)			
max_iter	Iterations to optimize weights	Int.	$[1, 1000]$
neurons	Number of neurons in the hidden layer	Int.	$[5, 1000]$
lr_init	Initial learning rate	Int.	$[1, 6]$
b1	First exponential decay rate	Real	$[10^{-7}, 1]$
b2	Second exponential decay rate	Real	$[10^{-7}, 1]$
Support Vector Machine (SVM)			
C	Regularization parameter	Real	$[0.1, 2]$
kernel	Kernel type to be used in the algorithm	Cat.	[linear, poly, rbf, sigmoid]
Decision Tree (DT)			
max_depth	Maximum depth of the tree. If 0, then *None* is used	Int.	$[0, 20]$
mss	Minimum number of samples required to split an internal node	Real	$[0, 0.99]$
msl	Minimum number of samples required to be at a leaf node	Int.	$[1, 10]$
max_f	Features in the best split	Cat.	[auto, sqrt, log2]
criterion	Measure the quality of a split	Cat.	[gini, entropy]

References

1. Ankenman, B., Nelson, B.L., Staum, J.: Stochastic kriging for simulation metamodeling. Oper. Res. **58**(2), 371–382 (2010). https://doi.org/10.1109/WSC.2008.4736089
2. Bergstra, J., Bardenet, R., Bengio, Y., Kégl, B.: Algorithms for hyper-parameter optimization. In: Advances in Neural Information Processing Systems, vol. 24 (2011)

3. Binois, M., Huang, J., Gramacy, R.B., Ludkovski, M.: Replication or exploration? Sequential design for stochastic simulation experiments. Technometrics **61**(1), 7–23 (2019). https://doi.org/10.1080/00401706.2018.1469433

4. Gonzalez, S.R., Jalali, H., Van Nieuwenhuyse, I.: A multiobjective stochastic simulation optimization algorithm. Eur. J. Oper. Res. **284**(1), 212–226 (2020). https://doi.org/10.1016/j.ejor.2019.12.014

5. Horn, D., Bischl, B.: Multi-objective parameter configuration of machine learning algorithms using model-based optimization. In: 2016 IEEE Symposium Series on Computational Intelligence (SSCI), pp. 1–8. IEEE (2016). https://doi.org/10.1109/SSCI.2016.7850221

6. Huband, S., Hingston, P., Barone, L., While, L.: A review of multiobjective test problems and a scalable test problem toolkit. IEEE Trans. Evol. Comput. **10**(5), 477–506 (2006)

7. Igel, C.: Multi-objective model selection for support vector machines. In: Coello Coello, C.A., Hernández Aguirre, A., Zitzler, E. (eds.) EMO 2005. LNCS, vol. 3410, pp. 534–546. Springer, Heidelberg (2005). https://doi.org/10.1007/978-3-540-31880-4_37

8. Jalali, H., Van Nieuwenhuyse, I., Picheny, V.: Comparison of kriging-based algorithms for simulation optimization with heterogeneous noise. Eur. J. Oper. Res. **261**(1), 279–301 (2017)

9. Knowles, J.: Parego: a hybrid algorithm with on-line landscape approximation for expensive multiobjective optimization problems. IEEE Trans. Evol. Comput. **10**(1), 50–66 (2006). https://doi.org/10.1109/TEVC.2005.851274

10. Li, L., Jamieson, K., DeSalvo, G., Rostamizadeh, A., Talwalkar, A.: Hyperband: a novel bandit-based approach to hyperparameter optimization. J. Mach. Learn. Res. **18**(1), 6765–6816 (2017)

11. Ozaki, Y., Tanigaki, Y., Watanabe, S., Onishi, M.: Multiobjective tree-structured parzen estimator for computationally expensive optimization problems. In: Proceedings of the 2020 Genetic and Evolutionary Computation Conference, pp. 533–541 (2020)

12. Quan, N., Yin, J., Ng, S.H., Lee, L.H.: Simulation optimization via kriging: a sequential search using expected improvement with computing budget constraints. IIE Trans. **45**(7), 763–780 (2013)

13. Snoek, J., Larochelle, H., Adams, R.P.: Practical Bayesian optimization of machine learning algorithms. In: Advances in Neural Information Processing Systems, vol. 25 (2012)

14. Williams, C.K., Rasmussen, C.E.: Gaussian Processes for Machine Learning, vol. 2. MIT Press, Cambridge (2006)

15. Zhan, D., Xing, H.: Expected improvement for expensive optimization: a review. J. Global Optim. **78**(3), 507–544 (2020). https://doi.org/10.1007/s10898-020-00923-x

Novel Optimization Techniques

A New Algorithm for Bi-objective Problems Based on Gradient Information

N. Aslimani[1]([✉]), E.-G. Talbi[1], and R. Ellaia[2]

[1] University of Lille, Lille, France
n.aslimani@yahoo.fr, el-ghazali.talbi@univ-lille.fr
[2] EMI, Mohammed V University of Rabat, Rabat, Morocco
ellaia@emi.ac.ma

Abstract. This paper presents a new approach for bi-objective optimization based on the exploitation of available gradient information.

The proposed algorithm integrates the MGDA method in a retroactive process which start with one of the anchor points and alternate descent and climb moves producing a zigzag movement around the FP.

The proposed approach can be considered as a path-relinking strategy, by generating a path-relinking in the objective space, between each pair of Pareto solutions.

The Retro-MGDA approach largely outperforms state-of-the-art and popular evolutionary algorithms both in terms of the quality of the obtained Pareto fronts (convergence, diversity and spread) and the search time.

Keywords: Pareto front · Bi-objective optimization · Gradient methods · MGDA algorithm · Anchor points

1 Introduction

Multi-objective optimization is a fundamental area of multi-criteria decision support, which many scientific and industrial community have to face. The resolution of a multi-objective optimization problem consists in determining the solution that best corresponds to the preferences of the decision maker among the good compromise solutions. In most real-world problems, it is not about optimizing just one criterion but rather optimizing multiple criteria and which are usually conflicting. In design problems, for example, more often than not a compromise has to be found between technological needs and cost goals. Multi-objective optimization therefore consists of optimizing several functions. The notion of single optimal solution in uni-objective optimization disappears for multi-objective optimization problems in favor of the notion of set of optimal Pareto solutions.

Most real optimization problems are described using several objectives or often contradictory criteria that must be optimized simultaneously. While, for problems including only one objective, the optimum sought is clearly defined,

© The Author(s), under exclusive license to Springer Nature Switzerland AG 2022
B. Dorronsoro et al. (Eds.): OLA 2022, CCIS 1684, pp. 49–61, 2022.
https://doi.org/10.1007/978-3-031-22039-5_5

this remains to be formalized for multi-objective optimization problems. Indeed, for a problem with two contradictory objectives, the optimal solution sought is a set of points corresponding to the best possible compromises to solve our problem.

Gradient based methods are suitable for solving single-objective optimization problems, with or without constraints. However, for multi-objective optimization, this is less well established. In order to solve MOPs using gradient based methods, a popular strategy is the weighted sum formulation (WSF), which consists in combining all the objectives into a single objective using a linear aggregation method. Multiple solutions can be obtained by varying the weight coefficients among the objective functions to hopefully obtain different Pareto solutions. Many other scalarization functions have been used in the literature, such as the normal boundary intersection (NBI) [1], normal constraint method (NC) [2], physical programming method (PP) [3], goal programming (GP) [4], ϵ-constraint method [5], and the directed search domain (DSD) [6,7]. Another gradient based strategies are used in memetic algorithms in which we hybridize multi-objective evolutionary algorithms (MOEAs) with local search strategies where gradient information is used to built a Pareto descent direction. For instance, Harada et al. [8] proposed a new gradient based local search method called the Pareto Descent Method, based on random selection of search directions among Pareto descents. Kim et al. [9] presented a directional operator to further enhance convergence of any MOEAs by introducing a local gradient search method to multi-objective global search algorithms. Recently, Lara et al. [10] compared hybrid methods with a new local search strategy without explicit gradient information and showed that using the gradient information was beneficial. Many other works propose pure gradient based methods for MOPs. In [11–13], a common descent direction is computed along all objectives. For instance, the multiple gradient descent algorithm was developed as an extension of the steepest descent method to MOPs [11]. Bosman have presented an analytical description of the entire set of descent directions, that he integrated in a new gradient based method for MOPs named CORL [14].

Without loss of generality, we assume that all objectives are to be minimized, then we consider a MOP of the form:

$$\min_{X \in S} \mathbf{F}(X) = (f_1(X), \cdots, f_m(X))^{\mathrm{T}} \tag{1}$$

where:

$f_k : \mathbb{R}^n \longrightarrow \mathbb{R}$, for $k \in \{1, \cdots, m\}$, denotes the objective functions, S is the design space: $S = \prod_{i=1}^{n} [l_i, u_i]$, X is the decision vector with n decision variables: $X = (x_1, \cdots, x_n) \in \mathbb{R}^n$.

For the convenience of later discussion, we introduce some basic concepts:

Definition 1. *Pareto dominance*: let $X = (x_1, x_2, ..., x_n)$ and $Y = (y_1, y_2, ..., y_n)$ be decision vectors (solutions). Solution X is said to dominate solution Y, denoted as $X \preceq Y$, if and only if :

$$\forall i \in [1, m] \; : \; f_i(X) \leq f_i(Y)) \wedge (\exists j \in [1, m] \; : \; f_j(X) < f_j(Y) \tag{2}$$

Definition 2. *Pareto optimal solution*: a solution X is Pareto optimal if it is not dominated by any other solution which means there is no other solution $Y \in S$ such that $Y \preceq X$.

Definition 3. *Pareto Set & Pareto front*: the set of all Pareto optimal solutions is called Pareto set (PS). The corresponding set of Pareto optimal objective vectors is called Pareto front (PF).

In this paper, we propose a new gradient based approach for bi-objective optimization problems using two complementary dynamics: The MGDA approach combined to a backtracking approach. Starting with one of the anchor points the proposed algorithm alternates the two approaches in a way that generates a zigzag move along the PF. In fact, after each MGDA descent we apply locally a reverse move to get a new point from which we operates again a new MGDA descent (toward the PF) and so on until meeting the second anchor point.

The remainder of this paper is organized as follows. Section 2 is a brief introduction to MGDA algorithm. Section 3 describes our proposed approach. Section 4 reports the computational results and finally, Sect. 5 presents the conclusions and future research perspectives.

2   The MGDA Algorithm

The Multi-Gradient Descent Algorithm (MGDA) was developed by Desideri [15] is an extension of the classical Gradient Descent Algorithm to multiple objectives. The basic idea of MGDA is to identify a direction common to all criteria along which the value of every objective improves: thus, for a multi-objective minimization problem we seek a descent direction common to all criteria. This algorithm is proved to converge to the Pareto Stationary solution.

The author defines the common descent vector as the unique element minimizing the norm in the convex hull $\overline{U}$ of the gradients of each objective, where $\overline{U}$ is defined for a given $x \in \mathbb{R}^n$ as:

$$\overline{U} = \{\omega \in \mathbb{R}^n, \omega = \sum_{i=1}^{n} \alpha_i \nabla f_i(x), \sum_{i=1}^{n} \alpha_i = 1\} \tag{3}$$

Indeed, since $\overline{U}$ is a closed, bounded and convex set associated in the affine space $\mathbb{R}^n$, then $\overline{U}$ admits a unique element of minimum norm, say δ. Two cases are possible:

- $\delta = 0$, and we say that x is a point of Pareto-stationarity, a necessary condition for Pareto-optimality.
- $\delta \neq 0$, and the directional derivatives of the objective functions satisfy the inequalities:

$$(\nabla f_i, \delta) \geq \|\omega\|^2 \tag{4}$$

Hence, $-\delta$ is a descent direction common to all the objective functions.

The MGDA procedure is given in the following algorithm

Algorithm 1: MGDA procedure

1: Initialize the design point $X = X^0$
2: Evaluate n objective $f_i(X)$; $\forall i = 1 \cdots n$; Compute the normalised gradient vectors
$\delta_i = \nabla f_i(X)/Si$; $\forall i = 1 \cdots n$;
3: Determine the minimal-norm element δ in the convex hull $\overline{U}$;
4: If $\delta = 0$ (or under a given tolerance) , stop ;
5: Else perform line search to determine the optimal step size ρ
6: Update design point X to $X - \rho\delta$

---

3 The Proposed Retro-MGDA Algorithm

3.1 Principle and Motivation

The Multiple Gradient Descent Method (MGDA) allows, starting from a giving point, to converge to a stationary Pareto point following a direction of common descent constructed by linear combination of the gradients of the objectives involved: The major drawback of this method (such as methods using the same principle) is that it can easily be trapped on a local Pareto front because of Pareto stationarity which exhausts the common direction of descent as we approach a pareto (local) front. Therefore this method is ineffective on Multi-objective problems presenting a multi-modality as illustrated in Fig. 1 (as for the gradient descent in the presence of multi-modality in the single-objectif context).

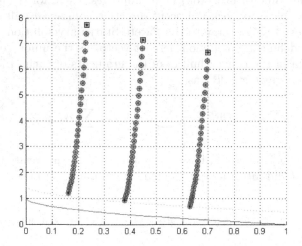

Fig. 1. Illustration of MGDA failure of MGDA to reach the PF of the multi-modal problem ZDT4

On the other hand, if the starting point is close to the Pareto Front then the MGDA succeeds in reaching the FP. Now if we move back locally from this point using a direction other than the direction of descent used by MGDA, we obtain a new starting point different from the first which allows, by applying the MGDA to reach a new point on the Front FP on from which we can start the process again as shon in Fig. 2.

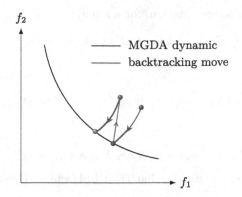

Fig. 2. Alternation of MGDA/backtracking dynamic

Thereby, it is possible to design a mechanism using the MGDA able to capture the Pareto Front with just a single starting point on the PF as illustrated in Fig. It remains to define the starting point, a natural choice is to consider one of the Anchor points is to use the predetermined directions (horizontal/vertical) built with available gradient information as illustrated in Fig. 3.

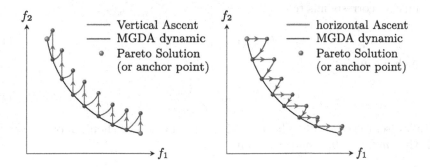

Fig. 3. Illustration of Retro-MGDA mechanism

3.2 Controlling Moves on the Objective Space by Using the Gradient Information

Consider a Point X such as $(\nabla f_1(X), \nabla f_2(X)) \neq (0,0)$.

Let $\eta_i = \|\nabla f_i(X)\|$, and consider the normalized gradients: $\delta_i = \nabla f_i(X)/\eta_i$. Let $\alpha = <\delta_1, \delta_2>$ and $t \in \mathbb{R}$. Consider the combined gradient direction δ_t built on the two gradients and defined as:

$$\delta_t = t\delta_1 + (1 - t)\delta_2. \tag{5}$$

Due to the linear approximation, for a small enough step λ:

$$F(X + \lambda\delta_t) - F(X) \simeq \lambda\delta_t \nabla F(X). \tag{6}$$

But:

$$\delta_t \nabla F(X) = \delta_1(\eta_1\delta_1, \eta_2\delta_2)^T = (\eta_1\delta_t\delta_1, \eta_2\delta_t\delta_2)^T. \tag{7}$$

Thus:

$$F(X + \lambda\delta_t) - F(X) \simeq \lambda(\eta_1\delta_t\delta_1, \eta_2\delta_t\delta_2)^T. \tag{8}$$

In the objective space, we say that the displacement is vertical if:

$$f_1(X + \lambda\delta_t) - f_1(X) = 0. \tag{9}$$

Likewise, we say that the displacement is horizontal if:

$$f_2(X + \lambda\delta_t) - f_2(X) = 0. \tag{10}$$

Hence, to get locally a vertical displacement, one should move along δ_{t_v} such that $\eta_1\delta_t\delta_1 = 0$ or $\delta_t\delta_1 = 0$. But:

$$\delta_t\delta_1 = (t\delta_1 + (1 - t)\delta_2)\delta_1 = t\delta_1^2 + (1 - t)\delta_1\delta_2 = t + (1 - t)\alpha. \tag{11}$$

Thus δ_{t_v} corresponds to

$$t_v + (1 - t_v)\alpha = 0, \tag{12}$$

which means:

$$t_v = \frac{\alpha}{\alpha - 1} = \frac{<\delta_1, \delta_2>}{<\delta_1, \delta_2> - 1}. \tag{13}$$

Likewise, to get locally a horizontal displacement, one should move along δ_{t_h} such that $\eta_1\delta_t\delta_1 = 0$ or $\delta_t\delta_1 = 0$. But:

$$\delta_t\delta_2 = (t\delta_1 + (1 - t)\delta_2)\delta_2 = t\delta_1\delta_2 + (1 - t)\delta_1^2 = t\alpha + 1 - t = t(\alpha - 1) + 1. \tag{14}$$

Thus δ_{t_h} corresponds to

$$1 + (1 - \alpha)t_h = 0, \tag{15}$$

which means:

$$t_h = \frac{1}{\alpha - 1} = \frac{1}{<\delta_1, \delta_2> - 1}. \tag{16}$$

Besides the two directions δ_h and δ_v, we can also consider the transverse direction $\delta_\perp = \frac{1}{2}(\delta_2 - \delta_1)$ which has the particularity of being orthogonal to the direction of descent $\delta = \frac{1}{2}(\delta_1 + \delta_2)$. Indeed:

$$\delta \cdot \delta_\perp = \frac{1}{2}(\delta_2 - \delta_1) \cdot \frac{1}{2}(\delta_1 + \delta_2) = \frac{1}{4}(\|\delta_2\|^2 - \|\delta_1\|^2) = 0. \tag{17}$$

3.3 Reto-MGDA Structure

Algorithm 2: RGDA pseudo code

1: **Input**: F, λ, (MGDA stepsize), α (backtracking stepsize)

2: **Output**: RND : Approximation of the PF

3: determine the anchor points X_1^*, X_2^*, which represent the solutions of single-objective problems $X_i^* = \underset{X \in S,\, l_i \leq x_i \leq u_i}{Argmin} f_i(X)$ (see Fig. 3).

4: Set $X = X_2^*$ (or $X = X_1^*$ with horizontal backtracking scenario)

5: Set $X_p = X$; $Y = F(X)$; $Ys = F(X_1^*)$

6: Set $\alpha = 0.005 \times \|F(X_2^*) - F(X_1^*)\|$;

7: **while** $Y_1 > Ys_1$ **do**

8:     Compute the normalized gradients of the two objectives: $\delta_i = \nabla f_i(X)/\eta_i$

9:     Compute the vertical direction using equations (5, 13): $\delta_t = t_v\delta_1 + (1 - t_v)\delta_2$.

10:     Normalize the vertical direction $\delta_t = \delta_t/\|\delta_t\|$

11:     Compute the transverse displacement according to : $X_d = X + \alpha\delta_\perp$
        ▷ We can also consider a variant with a vertical displacement involving δ_v

12:     Update the current solution X by applying the MGDA algorithm to X_d :
        $X_d = MGDA(X, \lambda)$

13:     Update the current objective: $Y = F(X)$

14: **end while**

---

4   Parameters Setting

The Retro-MGDA approach (or simply RGDA) can use any single objective algorithm to approximate the anchors points. In our experiments, we use a genetic algorithm (GA)[1] to handle this issue. The parameter setting in GA was as follows: the population size is 40, the crossover rate is $c = 0.9$, the mutation factor is $m = 0.1$, and the number of generation is 1000 for the first anchor point and 250 for the second anchor point. The step size of the MGDA move $\lambda = 0.01$. This parameter configuration was adopted for all the experiments. The algorithm have been run on each test problem for 10 times.

---

[1] Available in the yarpiz library https://www.yarpiz.com.

5   Experiments and Results

In order to evaluate the performance of the proposed RGDA algorithm, seven
test problems are selected from the literature: Zdt1, Zdt2, Zdt3, Zdt4, Zdt6,
Pol and Kur. These problems are covering different type of difficulties and are
selected to illustrate the capacity of the algorithm to handle diverse type of
Pareto fronts. In fact, all these test problems have different levels of complexity
in terms of convexity and continuity. For instance, the test problems Kur and
Zdt3 have disconnected Pareto fronts; Zdt4 has too many local optimal Pareto
solutions, whereas Zdt6 has non convex Pareto optimal front with low density
of solutions near Pareto front.

5.1   Performance Measures

Three performance measures were adopted in this study: the generational distance (GD) to evaluate the convergence, the Spacing (S) and the Spread (Δ) to
evaluate the diversity and cardinality.

- The convergence metric (GD) measure the extent of convergence to the true
  Pareto front. It is defined as:

$$GD = \frac{1}{N} \sum_{i=1}^{N} d_i, \tag{18}$$

where N is the number of solutions found and d_i is the Euclidean distance
between each solution and its nearest point in the true Pareto front.
- The Spread Δ, beside measuring the regularity of the obtained solutions, also
  quantifies the extent of spread in relation to the true Pareto front. The Spread
  is defined as:

$$\Delta = \frac{d_f + d_l + \sum_{i=1}^{N} |d_i - \overline{d}|}{d_f + d_l + d_i + (N-1)\overline{d}}. \tag{19}$$

where d_i is the Euclidean distance between two consecutive solutions in the
obtained set, d_f and d_l denotes the distance between the boundary solutions
of the true Pareto front and the extreme solutions in the set of obtained
solutions, $\overline{d}$ denotes the average of all distances d_i, $i = 1, 2, \cdots, N-1$ under
assumption of N obtained non-dominated solutions.
- The Spacing metric S indicates how the solutions of an obtained Pareto front
  are spaced with respect to each other. It is defined as:

$$S = \sqrt{\frac{1}{N} \sum_{i=1}^{N} (d_i - \overline{d})^2} \tag{20}$$

5.2   Numerical Results

The proposed algorithm RGDA is compared with three popular evolutionary algorithms: MOEA/D [16], NSGA-II [17] and PESA-II [18][2]

The obtained computational results are summarized in Table 1 in term of the mean and the standard deviation (Std) of the used metrics (GD, S, Δ) for 10 independent experiments, the average number of Pareto solutions found (NS), the average number of function evaluations (FEs), the average execution time in seconds (Time).

Table 1. Comparison of MOEA/D, NSGA-II, PESA-II and RGDA for some considered test problems.

Fct	Method	GD		S		Δ		NS	CPU(s)
		Mean	Std	Mean	Std	Mean	Std		
Zdt1	MOEA/D	2,81e−02	2,92e−02	2,42e−02	6,83e−03	9,95e−01	7,98e−02	100	86,67
	NSGA-II	9,17e−02	1,03e−02	2,17e−02	1,56e−03	7,95e−01	2,86e−02	100	105,08
	PESA-II	5,38e−02	5,31e−03	3,73e−01	2,68e−02	8,82e−01	6,01e−02	100	40,97
	RGDA	**8,36e−04**	**2,56e−07**	**9,16e−03**	**1,03e−06**	**7,52e−01**	**9,71e−07**	**116**	**9,58**
Zdt2	MOEA/D	1,32e−01	4,55e−02	2,66e−02	3,87e−03	1,13e+00	8,99e−03	100	61,61
	NSGA-II	1,49e−01	1,74e−02	1,65e−02	2,31e−03	9,10e−01	1,69e−02	100	113,18
	PESA-II	9,04e−02	3,42e−03	5,40e−01	4,61e−02	8,32e−01	2,73e−02	100	32,82
	RGDA	**2,65e−04**	**2,14e−05**	**3,34e−03**	**9,50e−05**	**7,34e−01**	**6,54e−06**	**115**	**10,61**
Zdt3	MOEA/D	2,08e−02	8,00e−03	6,01e−02	1,52e−02	1,19e+00	3,43e−02	100	82,50
	NSGA-II	6,27e−02	4,74e−03	3,75e−02	1,23e−02	8,25e−01	2,12e−02	100	112,91
	PESA-II	4,45e−02	3,42e−03	3,54e−01	3,22e−02	8,48e−01	1,04e−01	100	33,93
	RGDA	**4,70e−03**	**3,34e−04**	**5,38e−02**	**6,26e−04**	**9,62e−01**	**1,65e−03**	**33**	**8,48**
Zdt4	MOEA/D	8,28e−01	5,75e−01	1,70e−01	1,45e−01	1,09e+00	5,77e−02	100	62,65
	NSGA-II	4,46e−01	1,33e−01	3,03e−01	1,79e−01	8,85e−01	9,56e−02	100	129,75
	PESA-II	1,12e+01	5,16e−01	2,72e+01	4,86e+00	1,11e+00	5,16e−02	100	18,00
	RGDA	**1,03e−03**	**4,93e−04**	**1,11e−02**	**3,00e−03**	**7,52e−01**	**3,05e−04**	**60**	**8,54**
Pol	MOEA/D	5,40e−01	1,08e−01	1,35e+00	1,01e+00	1,25e+00	1,54e−01	100	91,11
	NSGA-II	2,48e+00	5,84e−02	1,75e+00	2,09e−03	9,72e−01	3,55e−03	100	123,86
	PESA-II	1,52e+01	6,02e+00	1,01e+01	1,16e+00	9,77e−01	9,85e−02	100	45,64
	RGDA	**1,48e−03**	**9,34e−06**	**6,03e−01**	**2,13e−02**	**9,59e−01**	**3,70e−08**	**22**	**5,24**
Zdt6	MOEA/D	4,22e−01	1,18e−01	4,63e−02	2,02e−02	1,09e+00	5,98e−02	100	57,86
	NSGA-II	3,08e−01	2,38e−02	1,57e−01	4,15e−02	8,73e−01	2,93e−02	100	121,89
	PESA-II	4,42e−01	4,69e−03	2,00e+00	6,31e−01	1,03e+00	4,97e−02	100	29,17
	RGDA	**7,83e−03**	**6,31e−03**	**1,04e−01**	**4,14e−03**	**7,20e−01**	**2,05e−02**	**35**	**5,65**
Kur	MOEA/D	5,51e−03	5,51e−03	6,76e−01	4,35e−01	1,42e+00	1,80e−01	100	80,83
	NSGA-II	5,31e−04	2,64e−05	1,22e−01	4,15e−03	4,10e−01	3,10e−02	100	117,86
	PESA-II	1,78e−01	1,28e−02	4,15e+00	4,86e−01	9,04e−01	5,75e−02	100	33,76
	RGDA	**3,71e−03**	**3,05e−04**	**1,50e−01**	**8,56e−03**	**9,03e−01**	**9,92e−06**	**24**	**12,02**

By analyzing the GD metric statistics, we can see that the proposed RGDA is well converging to the true Pareto front for all tested problems.

Furthermore, the Spacing and Spread measures indicate that the proposed RGDA has the ability of generating uniform and diverse solutions. Except for

---

[2] MATLAB implementation obtained for the yarpiz library available at https://www.yarpiz.com.

the Zdt6 problem whose PF presents folds (Pareto many-to-one problems) [19] which leads to the presence of several points with a zero gradient according to one of the two objectives. In this case RGDA, can only get a partial capture of the FP. To prevent this problem we have been led to slightly shift the points where this occurs, which has partially overcome this problem. Indeed, despite this precaution, it happens that the RGDA does not capture the whole of the FP for this problem (see Fig. 4(f)).

Note that all the simulations for the considered problems at worst did not exceed 70,000 FEs of which 50,000 FEs were reserved for the resolution of the

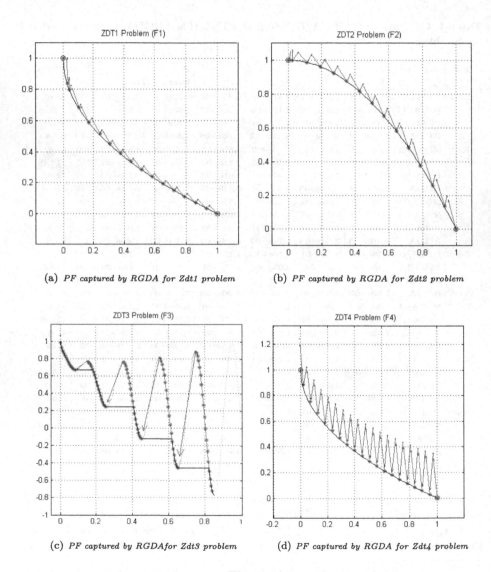

(a) *PF captured by RGDA for Zdt1 problem*   (b) *PF captured by RGDA for Zdt2 problem*

(c) *PF captured by RGDAfor Zdt3 problem*   (d) *PF captured by RGDA for Zdt4 problem*

Fig. 4.

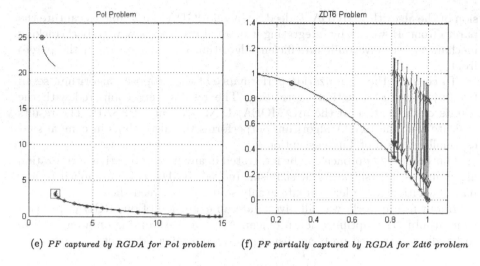

(e) *PF captured by RGDA for Pol problem*   (f) *PF partially captured by RGDA for Zdt6 problem*

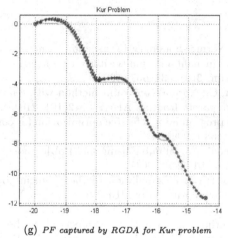

(g) *PF captured by RGDA for Kur problem*

Fig. 4. (*continued*)

anchors points. However, the number of solutions obtained strongly depends on the step of the retreat which was previously set at 0.5% of the difference D_a between the two anchor solutions. Thus, if we decide to reduce this step for example to 0.1% of D_a, we can easily multiply the number of solutions obtained while respecting the limit of the FEs imposed for the comparison.

6 Conclusion

In this paper, we have proposed a new approach for bi-objective problems, termed as RGDA, which adopts the MGDA approach combined to a backtracking approach in order. The proposed algorithm can be considered as an extension of the algorithm MGDA. Indeed, by itself MGDA is not able to capture the pareto front. However, by integrating it in an alternative way with backtracking mechanism, we have built an efficient algorithm for the capture of the pareto front.

To consider the optimization performance of the proposed algorithm, seven two-objective test functions were used. The results were compared with the results of three state of the art: MOEA/d, NSGAII and PESAII. The results indicate that the RGDA algorithm outperforms the algorithms in comparisons especially in term of Time execution.

Our proposed approach can be integrated in any multi-objective optimization algorithm as intensification mechanism. Indeed, RGDA can be applied to any pair of solutions in order generate evenly solutions between them.

In our future work, we will investigate an extension of our RGDA approach to multi-objective optimization problem. Also, we are working on a mechanism to handle constrained problems using projected gradient descent.

References

1. Das, I., Dennis, J.: Normal-boundary intersection: a new method for generating pareto optimal point in nonlinear multicriteria optimization problems. SIAM J. Optim. J. Glob. Optim. **3**(8), 631–657 (1998)
2. Messac, A., Mattson, C.: Normal constraint method with guarantee of even representation of complete pareto frontier. AIAA J. **42**(10), 2101–2111 (2004)
3. Messac, A.: Physical programming-effective optimization for computational design. AIAA J. **34**(1), 149–158 (1996)
4. Deb, K.: Nonlinear goal programming using multi-objective genetic algorithms. J. Oper. Res. Soc. **3**(52), 291–302 (2001)
5. Haimes, Y.: On a bicriterion formulation of the problems of integrated system identification and system optimization. IEEE Trans. Syst. Man Cybern. **1**(3), 296–297 (1971)
6. Utyuzhnikov, S., Fantini, P., Guenov, M.: Numerical method for generating the entire pareto frontier in multiobjective optimization. In: Proceedings of EURO-GEN, pp. 12–14 (2005)
7. Utyuzhnikov, S., Fantini, P., Guenov, M.: A method for generating a well-distributed pareto set in nonlinear multiobjective optimization. J. Comput. Appl. Math. **223**(2), 820–841 (2009)
8. Harada, K., Sakuma, J., Kobayashi, S.: Local search for multiobjective function optimization: pareto descent method. In: Genetic and Evolutionary Computation Conference GECCO, pp. 659–666, 755–762 (2006)
9. Kim, H.J., Liou, M.S.: New multi-objective genetic algorithms for diversity and convergence enhancement. AIAA J. (2009)
10. Lara, A., Sanchez, G., Coello, C., Schutze, O.: HCS: a new local search strategy for memetic multiobjective evolutionary algorithms. IEEE Trans. Evol. Comput. **14**(1), 112–132 (2009)
11. Desideri, J.-A.: Mutiple-gradient descent algorithm for multiobjective optimization. In: Berhardsteiner, J. (ed.) European Congress on Computational Methods in Applied Sciences and Engineering (ECCOMAS) (2012)

12. Fliege, J., Svaiter, B.F.: Steepest descent methods for multicriteria optimization. Math. Methods Oper. Res. **51**(3), 479–494 (2000)
13. Schäffler, S., Schultz, R., Weinzierl, K.: Stochastic method for the solution of unconstrained vector optimization problems. J. Optim. Theory Appl. **114**(1), 209–222 (2002)
14. Bosman, P.: On gradients and hybrid evolutionary algorithms. IEEE Trans. Evol. Comput. **1**, 51–69 (2012)
15. Désidéri, J.-A.: Multiple-gradient descent algorithm (MGDA) for multiobjective optimization. Comptes Rendus Mathematique **350**(5–6), 313–318 (2012)
16. Zhang, Q., Li, H.: MOEA/D: a multiobjective evolutionary algorithm based on decomposition. IEEE Trans. Evol. Comput. **11**(6), 712–731 (2007)
17. Deb, K., Pratap, A., Agarwal, S., Meyarivan, T.: A fast and elitist multiobjective genetic algorithm: NSGA-II. IEEE Trans. Evol. Comput. **6**(2), 182–197 (2002)
18. Corne, D., Jerram, N., Knowles, J., Oates, M.: PESA-II: region-based selection in evolutionary multiobjective optimization. In: Proceedings of the 3rd Annual Conference on Genetic and Evolutionary Computation, pp. 283–290 (2001)
19. Huband, S., Hingston, P., Barone, L., While, L.: A review of multiobjective test problems and a scalable test problem toolkit. IEEE Trans. Evol. Comput. **10**(5), 477–506 (2006)

Adaptive Continuous Multi-objective Optimization Using Cooperative Agents

Quentin Pouvreau[1,2]($\boxtimes$)(iD), Jean-Pierre Georgé[1](iD), Carole Bernon[1](iD), and Sébastien Maignan[1]

[1] IRIT, Université de Toulouse, CNRS, Toulouse INP, UT3, Toulouse, France
{quentin.pouvreau,jean-pierre.george,carole.bernon,sebastien.maignan}@irit.fr
[2] ISP System, Vic-en-Bigorre, France
https://www.irit.fr/, https://www.isp-system.fr/

Abstract. Real-world optimization of complex products (*e.g.*, planes, jet engines) is a hard problem because of huge multi-objective and multi-constrained search-spaces, in which many variables are to be adjusted while each adjustment essentially impacts the whole system. Since the components of such systems are manufactured and output values are obtained with sensors, these systems are subject to imperfections and noise. Perfect *digital twins* are therefore impossible. Furthermore simulating with sufficient details is costly in resources, and the relevance of Population-based optimization approaches, where each individual is a whole solution to be evaluated, is severely put in question. We propose to tackle the problem with a Multi-Agent System (MAS) modeling and optimization approach that has two major strengths: 1) a natural representation where each agent is a variable of the problem and is perceiving and interacting through the real-world topology of the problem, 2) a cooperative solving process where the agents continuously adapt to feedback, that can be interacted with, can be observed, where the problem can be modified on-the-fly, that is able to directly control these variables on a real-world product while taking into account the specifics of the components. We illustrate and validate this approach in the *Photonics* domain, where a light beam has to follow a path through several optical components so as to be transformed, modulated, amplified, etc., at the end of which sensors give feedback on several metrics that are to be optimized. Robotic arms have to adjust the 6-axis positioning of the components and are controlled by the Adaptive MAS we developed.

Keywords: Continuous optimization · Multi-objective optimization · Adaptive multi-agent systems · Robotics control · Photonics

1 Problem Statement and Positioning

This study mainly concerns optimization problems from real-world applications, especially robotics control command based on sensor feedback. These applications go from system configuration based on test bench feedback to real-time

B. Dorronsoro et al. (Eds.): OLA 2022, CCIS 1684, pp. 62–73, 2022.
https://doi.org/10.1007/978-3-031-22039-5_6

feedback control of an automated system. In the first case we want to optimize the response to an input. In the second case we mainly want to minimize the distance between sensor feedback and objectives by positioning, orienting, aligning one or more components. We define a robot, an actuator or any automated system as a composition of one or more axes, which are associated with the degrees of freedom of the system. Such problems have a wide variety of external constraints like limited resolution time or limited number of moves, predetermined components positions to respect, etc. We intend to develop an optimization system able to adapt to multiple robotics applications. The application domain we focus on is the photonics domain as illustrated in Fig. 1.

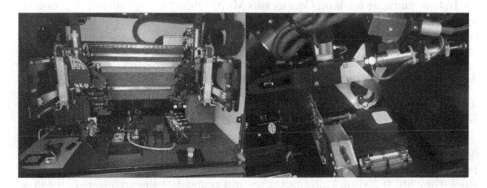

Fig. 1. Example of a real photonics application we are working on (Credits ISP System)

1.1 Search Space Topology

Optimization problems are divided into domains depending on their search space topology. Search space dimensions are defined by decision variables, and potentially limited by hard constraints. For example the number of robots and the number of axes per robot increase the dimensions. On the other hand, limits on a component assembly reduce the possibilities on one or more axes, so one or more dimensions get constrained in the corresponding search space. These constraints, alongside one or more expert-given objectives, most of the time antinomic, make appear a Pareto front in the search space. As constraints can generally be transposed into objectives dealing with the distance to a threshold (the further away from the threshold, the better, or defining a cost regarding an acceptable violation), these problems can be defined as Multi-Objective Optimization (MOO) problems, also called Pareto optimization problems.

Another factor of dimensioning of the search space is the decision variable domain. Depending on whether the domain is discrete or continuous, optimization problems are combinatorial or continuous. Since the precision of robotic systems is continuously increasing, robots positioning problems can be considered as continuous optimization problems.

Indeed, the axis resolution, meaning the minimal step with accuracy guarantee, is often very small compared to the range of values it can achieve. Furthermore, axes resolutions, ranges and other robots features can change from an application to another, changing therefore the decision variable domains.

A continuous problem has a potentially infinite number of solutions when a combinatorial problem has only all possible combinations of its discrete variables. This gap can have a significant influence in terms of calculation cost. However, those properties do not mean that combinatorial problems are trivial and continuous problems are not. It means that search space topology is quite different from a category to another so the employed method might not have the same results.

In this study, we focus on Continuous Multi-Objective Optimization problems and we consider that reliable problems with only one objective are a particular subdivision of these problems and can be processed in the same way.

1.2   Multi-objective Optimization Approaches

This section introduces the main approaches used in the optimization domain before focusing on the most suited for our concern.

Analytical approaches are the most accurate but they are time-consuming and may not also be suitable for large-scale optimization problems. Arithmetic programming approaches on the contrary have fast computation performances since they are based on simplifications and sequential linearizations. However, they are very weak in handling multi-objective nonlinear problems and may converge to local optima, non-necessarily satisfying enough [16].

Meta-heuristic optimization algorithms are extensively used in solving Multi-Objective Optimization problems since they can find multiple optimal solutions in a single run, and improve the ratio between accuracy and computational cost. They are problem-independent optimization techniques which provide, if not optimal, at least satisfying solutions by stochastically exploring and exploiting search spaces iteratively [17]. The following sections focus on these algorithms.

Population-Based Heuristics are a large part of the state of the art in Meta-heuristic Optimization Algorithms. The main principle is to simultaneously handle a population of solutions spread randomly or not in the search space. The population can evolve and select the best solutions iteratively, or converge to an optimum following a set of influence rules. These algorithms can also be used in hybrid solutions alongside more classical algorithms like simulated annealing [1] to balance their weaknesses.

It is difficult to be exhaustive about all works in this domain. For instance, Genetic Algorithms [9] and Particle Swarm Optimization Algorithms [12] have together more than 3000 publications per year [18].

A major limitation of Population-Based approaches is their potential computational cost. Such algorithms need to evaluate a relatively large number of candidates in order to create a good population of solutions. Computing all the candidate solutions can be prohibitive for computationally expensive problems.

The type of applications we are studying here requires an adaptation each time we get a sensor feedback. As a result, an evolution process would require a huge amount of resources. As each new candidate in the population needs to be evaluated, the robot needs to reconfigure the whole experimental setting for each proposed configuration. An adaptive algorithm modifying and proposing for testing a unique configuration at each feedback is a more suitable strategy (i.e. a resolution process forming a single trajectory in the search space).

Moreover, when scaling up the number of objectives, the Pareto-dominance relation essentially loses the ability to distinguish desirable solutions, since nearly all population members are non-dominated at an early stage of the search. In fact, a large majority of the usual Population-Based methods (evolutionary algorithms, swarm intelligence) have been shown to degrade when the number of objectives grows beyond three, and moreover beyond eight [10]. This particularity explains why a part of the literature about these approaches focuses on Many-Objective Optimization, that is Multi-Objective Optimization problems with more than three objectives [13]. The need for this category of problems to have more relevant indicators than Pareto-dominance makes the Population-Based Heuristics to specify additional calculation for solution comparison.

The types of problems we are interested in require an optimization system able to converge without maintaining and computing a large number of solutions, especially when solution comparison becomes non trivial. Furthermore, we need a decentralized real-time control of robots arms to actually optimize the real world system being processed, taking into account errors and noise.

Multi-Agent Systems (MAS) are a decentralized approach, based on self-organisation mechanisms [22], where the calculation task is distributed over *agents* which are virtual or physical autonomous entities [19].

Each agent has only a local point of view of the problem it is solving, corresponding to a local objective function. The global objective function of a problem is then the sum of all these local functions. This particularity enables to easily distribute calculation tasks in the resolution process and consequently reduces computational costs. That is why multi-agent approaches are preferred where centralized approaches have limited flexibility and scalability.

Multi-Agent Systems are used in a wide variety of theoretical [15] and real-world application domains of distributed optimization [21]: classification optimization algorithms [3], power systems [7], complex networks and IoT [5], smart manufacturing [2] or multi-robot systems [20].

Distributed Constraint Optimization Problems (DCOPs) are a well-known class of combinatorial optimization problems prevalent in Multi-Agent Systems [4]. DCOP is a model originally developed under the assumption that each agent controls exactly one variable.

This model was designed for a specific type of problems where the difficulty resides in the combination of multiple constraints. These problems are supposed to be easily decomposable into several cost functions, where the cost values associated with the variables states are supposed to be known. This major assumption does not stand for complex continuous optimization problems, where the complexity of the models and their interdependencies cause this information to be unavailable in most cases. It is important to remark that some works tried to extend DCOP model to continuous optimization problems but the state of art about those works remains scattered for now [8].

2   AMAS Theory for Optimization

As seen before, a MAS is a problem-independent solution making it possible to have a natural representation where each agent is a variable of an MOO problem. These agents, which perceive their environment and interact, are also a means to continuously adapt to real-world feedback and provide a "solution" anytime especially when the problem can be modified on-the-fly.

We propose to adopt the Adaptive Multi-Agent Systems (AMAS) theory where cooperation [6] is the engine that drives the adaptation of an agent and the emergence of a global functionality. This cooperation relies on three mechanisms: an agent may adjust its internal state to modify its behavior (tuning), may modify the way it interacts with its neighborhood (reorganization) or may create other agents or self-suppress when there is no other agent to produce a functionality or when a functionality is useless (evolution).

2.1   Natural Domain Modeling

As we previously stated, when solving complex continuous problems existing techniques usually require a transformation of the initial formulation, in order to satisfy some requirements for the technique to be applied. Beside the fact that correctly applying these changes can be a demanding task for the designers, imposing such modifications changes the problem beyond its original, natural meaning. What we propose here is an agent-based modeling where the original structure/meaning of the problem, is preserved. Indeed it represents the formulation which is the most natural and easiest for the expert to manipulate. We call this modeling *Natural Domain Modeling for Optimization* (NDMO) [11].

In order to represents the elements of a generic continuous optimization model, we identified five classes of interacting entities: *models, design variables, outputs, constraints* and *objectives*. Briefly: given the values of the design variables, certain models will calculated output values, which will enable other models to calculate other outputs and so on, until constraints and objectives can be calculated, in a sort of calculus network. In general, three elements need to be *agentified*: the design variables (because they need to be optimised and that constitutes the solving process), the constraints and objectives. The last two are there to model the requirements or statements of the problem, i.e. what the solving process has to achieve.

To this end, we use a mechanism based on a specific measure called *criticality*. This measure represents the state of dissatisfaction of the agent regarding its local goal. Each agent is in charge of estimating its own criticality and providing it to the other agents. The role of this measure is to aggregate into a single comparable value all the relevant indicators regarding the state of the agent. Having a single indicator of the state of the agent is interesting as it simplifies the reasoning of the agents. However the system designer has the difficult task to provide the agents which adequate means to calculate their criticality.

2.2   Agent Internal State and Behavior

Our system is a work-in-progress implementation based on the AMAK Framework [14]. A main system representing the AMAS handles an environment representation and a collection of agents interacting with the environment. Each agent controls a parameter of the system, consequently a decision variable of the problem. The environment and the set of agents execute an iteration to update their state one at a time. The agents iteration order is randomly updated at the beginning of each cycle. All the agents have the same three-phase algorithm:

- Perception phase: the agent gets an observation of the environment, that is a set of *criticalities* calculated from the distance to the objective.
- Decision phase: the agent processes its new data and follows a decision tree to adjust its internal state.
- Action phase: the agent acts following its decision by changing its parameter.

The agent decision phase aims at increasing or decreasing the value of the decision variable it is responsible for (a predefined variation step depending on its characteristics), and is thus at the core of the process. Except in one case that we will explain below, a decision is always repeated a stochastically chosen number of iterations, from one to ten. This *momentum* mechanism allows to desynchronize the agents decisions to prevent them from being trapped in what we call *non cooperative synchronisations* (basically when two agents try to "help" at the same time, thus hindering each other).

When starting the resolution, each agent has only a set of criticalities given by the last environment update. Since it does not have any idea of which action is the best, its first decision is to act randomly. If the agent observes the criticalities decreased beyond a configurable threshold, it will repeat its decision. It is generally useful to converge fast when the current region of the search space is relatively regular. On the contrary, if the criticalities increased beyond the same threshold, the agent will make only one step in the opposite way. These two rules make a first decision process we can qualify as reactive. When it is not possible to reactively spot an adequate decision, the agent will process data it registered in the last perception phases. This more cognitive process consists in interpolating the variation of the criticalities according to its own value. The goal is to identify a region (plus or minus) where the integrative is the lower. It appears that the momentum mechanism is also useful to this decision process since it made the agent do what we might call a stochastic scanning of its local area. In the rare case the second decision process fails to give a decision, the agent acts randomly.

3   Photonics Problem Modeling and Implementation

The environment of the AMAS has to update the system state, apply the changes from the AMAS and calculate the input variables used by the agents to take their decisions. The simulator we developed is shown in Fig. 2.

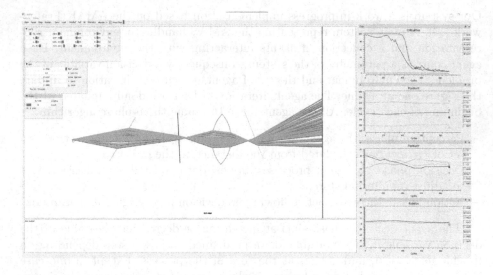

Fig. 2. Screenshot of the simulator at runtime

The system is a 2D-world composed of a light source, several lenses (L_i with i in $[1, N]$) and a screen. The light source emits a number of rays (R_j with j in $[1, M]$) in a conic shape. If a ray intersects with a lens it is refracted using Snell's law of refraction i.e. $n1.sin(\theta_1) = n2.sin(\theta_2)$ (with each θ as the angle measured from the normal of the boundary, and n as the refractive index of the respective medium). So assuming a ray goes through all lenses in the system (which is the desired state) we have a mathematical suite of operations applied to its position and orientation $R_j(pos_{j,i}, \theta_{j,i}) = L_i(R_j(pos_{j,i-1}, \theta_{j,i-1}))$ with i in $[1, n]$ and j in $[1, m]$.

Test cases for the AMAS are generated so that all rays pass through all lenses as follows: a set of lenses with various characteristics (thin or cylindrical, refraction index, focal length) are randomly placed on the axis between the source and the screen (X). A set of rays is generated parallel to the X axis so that each ray goes through all lenses and reaches the screen. Repeatedly, the lenses are shifted and rotated randomly as well as the direction of the rays, while keeping all rays going through all lenses. After a number of cycles, the state of the system is set as the reference to reach for the AMAS. Then the lenses are randomly placed and rotated. This new state is used as a starting point for the AMAS to work with.

A lens L_i is represented by its type (thin or cylindrical), its position $P_i(x, y)$ and rotation angle T_i, its refraction index n_i and focal length F_i. For cylindrical lenses the radius of each face are also necessary $R1_i$ and $R2_i$. From these parameters only P_i and T_i can change during the run and are controlled by the AMAS.

A ray is a more complex structure since it is represented by a path. A path is an ordered list of positions and directions describing the intersection points with the various lenses it goes through $\{p_{j,k}(x, y)\}$ with k the index of the intersection, and the direction of the ray at these points represented as an angle with X $\{ang_{j,k}\}$.

Rays are not directly known by the AMAS. Only the last position and direction of each ray (when it reaches the screen) is used to derive information to send to the AMAS as a feedback to its actions.

The derived information can be the results of various computations. The most straightforward is to form a set of M differences between current rays and reference rays: $\{|p_{j,last}(y) - p_{j,last}^{ref}(y)|, |ang_{j,last} - ang_{j,last}^{ref}|\}$. This gives the AMAS quite a lot of precise information which is not often readily available in real life systems.

The second type of derived information are root mean square deviations ($rmsd$) of the positions and angles: $R_{pos} = \sqrt{(\sum((p_{j,last}(y) - p_{j,last}^{ref}(y))^2)/M)}$ and $R_{ang} = \sqrt{(\sum((ang_{j,last} - ang_{j,last}^{ref})^2)/M)}$.

These two values are more representative of what can be perceived on a real system like the global intensity on the screen.

For each of the experiments presented hereafter a set of parameters are given which represents the setup of the run: first, the sequence of lenses, from source to screen, present in the system with C for a cylindrical lens and T for a thin lens. Then the number of rays, the percentage of maximum step for lenses moves and the type of information sent to the AMAS (full or $rmsd$). So for an experiment with 3 lenses, 10 rays, 50% of maximum step and using the $rmsd$, this set would be $\{CTC, 10, 50\%, rmsd\}$.

4   Experiments

First we checked the ability to manage different types of lenses, as for instance, a cylindrical lens has more complex interactions with rays than a thin lens.

We generated two experiments $\{T, 100, 10\%, rmsd\}$ and $\{C, 100, 10\%, rmsd\}$ (Fig. 3 and Fig. 4) in which all the other parameters were equal. As visible on the graphs, the evolution of criticality is more chaotic with a cylindrical lens.

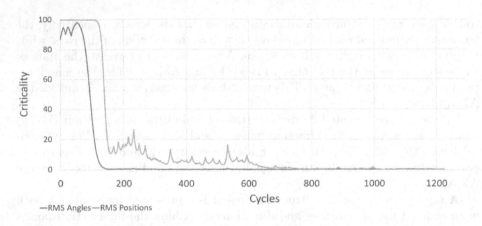

Fig. 3. Resolution with one thin lens $\{T, 100, 10\%, rmsd\}$

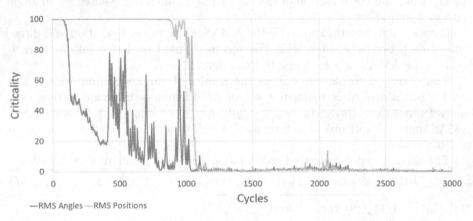

Fig. 4. Resolution with one cylindrical lens $\{C, 100, 10\%, rmsd\}$

We remark in the one lens experiments that curves make some peaks repetitively. These are the consequence of the momentum mechanism seen in Sect. 2.2 and do not impact the convergence that remains globally continuous.

The other experiments (Figs. 5 and 6) show that the system seems to be scalable in terms of number of decision variables. In these cases, the peaks mentioned earlier disappeared. The results of the moves of each axis agent are more softened as the number of interactions between rays and optical surfaces grows.

This proof of concept shows promising results: the resolution process succeeds and no divergence from a satisfying area of the search space has been observed. However, some adjustments that have to be explored yet could greatly improve the resolution process. The difficulties encountered are mainly due to the problem itself: almost all positioning values impact all criticalities, and in a non-linear way. The parameters consequently are strongly interconnected: the current position of one axis agent moves the target position of one or more others. Therefore,

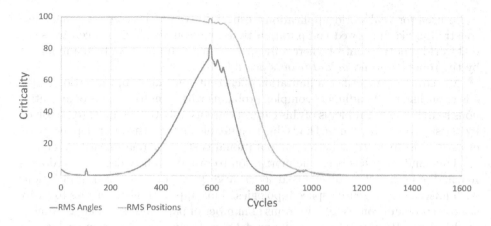

Fig. 5. Resolution with two lenses $\{CT, 100, 1\%, rmsd\}$

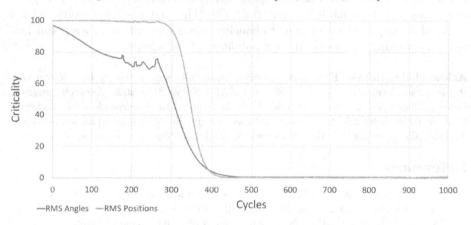

Fig. 6. Resolution with three lenses $\{TTT, 100, 1\%, rmsd\}$

the system is chaotic and the agents can collectively hinder themselves. At this point the optimization problem becomes a cooperation problem.

It has to be noted that the examples presented here are voluntary more difficult for the AMAS than a real case, where the starting positions are nearer from the optimum. Starting further away lets us test how the system behaves while crossing the vast search space of the problem. In this way we can observe that it does not suffer divergence from any achievement it has already made. Further work will be done to optimise the number of cycles needed to bring the criticalities down near zero.

5   Conclusion and Perspectives

State of the art in multi-objective optimization is dominated by Population-Based and DCOP approaches. However, it can be difficult for those methods to be used for real-world applications, especially when the context brings new constraints like narrowed computation time or resources. This is even worse in real world robotics control where only one "current solution" can be manipulated by the robots, and no *digital twin* is possible.

Moreover, we identified a limitation of current continuous optimization methods regarding the handling of complex problems with a multi-dimensional continuous search space. Problems of this category are usually too complex to be solved by classical optimization methods due to multiple factors: the inter-dependencies of their objectives, their heavy computational cost, their non-linearities, etc.

This limitation has been the motivation to propose a new decentralized approach. We designed a solver fitted for a large panel of real-world applications with miscellaneous search space topologies. This approach also permits to easily scale up problem complexity, in terms of number of parameters as well as number of objectives. Its aim is to naturally model a real-world optimization problem as a cooperative resolution problem and to satisfy as much as possible expert given objectives at a reasonable computation cost. The first results obtained with a proof of concept are promising. Enhancing the optimization process will now rely on enriching the cooperation capabilities of the agents.

Acknowledgements. This work has been financially supported by the *Région Occitanie* (www.laregion.fr) as part of the READYNOV 2019–2020 research program. Quentin Pouvreau has been co-funded by the *Association nationale de la recherche et de la technologie (ANRT)* (www.anrt.asso.fr) and by *ISP System*(www.isp-system.fr), who also provided the test cases and expert knowledge on photonics.

References

1. Assad, A., Deep, K.: A hybrid harmony search and simulated annealing algorithm for continuous optimization. Inf. Sci. **450**, 246–266 (2018)
2. Bendul, J.C., Blunck, H.: The design space of production planning and control for Industry 4.0. Comput. Ind. **105**, 260–272 (2019)
3. Couellan, N., Jan, S., Jorquera, T., Georgé, J.P.: Self-adaptive support vector machine: a multi-agent optimization perspective. Expert Syst. Appl. **42**(9), 4284–4298 (2015)
4. Fioretto, F., Pontelli, E., Yeoh, W.: Distributed constraint optimization problems and applications: a survey. J. Artif. Intell. Res. **61**, 623–698 (2018)
5. Fortino, G., Russo, W., Savaglio, C., Shen, W., Zhou, M.: Agent-oriented cooperative smart objects: from IoT system design to implementation. IEEE Trans. Syst. Man Cybern. Syst. **48**(11), 1939–1956 (2017)
6. Georgé, J.P., Gleizes, M.P., Camps, V.: Cooperation. In: Di Marzo Serugendo, G., Gleizes, M.P., Karageorgos, A. (eds.) Self-organising Software, pp. 193–226. Springer, Heidelberg (2011). https://doi.org/10.1007/978-3-642-17348-6_9
7. González-Briones, A., De La Prieta, F., Mohamad, M.S., Omatu, S., Corchado, J.M.: Multi-agent systems applications in energy optimization problems: a state-of-the-art review. Energies **11**(8), 1928 (2018)
8. Hoang, K.D., Yeoh, W., Yokoo, M., Rabinovich, Z.: New algorithms for continuous distributed constraint optimization problems. In: Proceedings of the 19th International Conference on Autonomous Agents and MultiAgent Systems, pp. 502–510 (2020)

9. Holland, J.H.: Adaptation in Natural and Artificial Systems: An Introductory Analysis with Applications to Biology, Control, and Artificial Intelligence. MIT Press, Cambridge (1992)
10. Ishibuchi, H., Tsukamoto, N., Nojima, Y.: Evolutionary many-objective optimization: a short review. In: 2008 IEEE Congress on Evolutionary Computation (IEEE World Congress on Computational Intelligence), pp. 2419–2426. IEEE (2008)
11. Jorquera, T., Georgé, J.P., Gleizes, M.P., Régis, C.: A natural formalism and a multiagent algorithm for integrative multidisciplinary design optimization. In: IEEE/WIC/ACM International Conference on Intelligent Agent Technology - IAT, pp. 146–154. Atlanta (2013)
12. Kennedy, J., Eberhart, R.: Particle swarm optimization. In: Proceedings of ICNN 1995-International Conference on Neural Networks, vol. 4, pp. 1942–1948. IEEE (1995)
13. Maltese, J., Ombuki-Berman, B.M., Engelbrecht, A.P.: A scalability study of many-objective optimization algorithms. IEEE Trans. Evol. Comput. $22(1)$, 79–96 (2016)
14. Perles, A., Crasnier, F., Georgé, J.-P.: AMAK - a framework for developing robust and open adaptive multi-agent systems. In: Bajo, J., et al. (eds.) PAAMS 2018. CCIS, vol. 887, pp. 468–479. Springer, Cham (2018). https://doi.org/10.1007/978-3-319-94779-2_40
15. Sghir, I., Hao, J.K., Jaafar, I.B., Ghédira, K.: A multi-agent based optimization method applied to the quadratic assignment problem. Expert Syst. Appl. $42(23)$, 9252–9262 (2015)
16. Shaheen, A.M., Spea, S.R., Farrag, S.M., Abido, M.A.: A review of meta-heuristic algorithms for reactive power planning problem. Ain Shams Eng. J. $9(2)$, 215–231 (2018)
17. Sharma, M., Kaur, P.: A comprehensive analysis of nature-inspired meta-heuristic techniques for feature selection problem. Arch. Comput. Methods Eng. $28(3)$, 1103–1127 (2020). https://doi.org/10.1007/s11831-020-09412-6
18. Wang, Z., Qin, C., Wan, B., Song, W.W.: A comparative study of common nature-inspired algorithms for continuous function optimization. Entropy $23(7)$, 874 (2021)
19. Weiß, G.: Multiagent Systems: A Modern Approach to Distributed Artificial Systems. MIT Press, Cambridge (1999)
20. Yan, Z., Jouandeau, N., Cherif, A.A.: A survey and analysis of multi-robot coordination. Int. J. Adv. Rob. Syst. $10(12)$, 399 (2013)
21. Yang, T., et al.: A survey of distributed optimization. Annu. Rev. Control. 47, 278–305 (2019)
22. Ye, D., Zhang, M., Vasilakos, A.V.: A survey of self-organization mechanisms in multiagent systems. IEEE Trans. Syst. Man Cybern. Syst. $47(3)$, 441–461 (2016)

Integer Linear Programming Reformulations for the Linear Ordering Problem

Nicolas Dupin$^{(\boxtimes)}$ (iD)

Univ Angers, LERIA, SFR MATHSTIC, 49000 Angers, France
`nicolas.dupin@univ-angers.fr`

Abstract. This article studies the linear ordering problem, with applications in social choice theory and databases for biological datasets. Integer Linear Programming (ILP) formulations are available for linear ordering and some extensions. ILP reformulations are proposed, showing relations with the Asymmetric Travel Salesman Problem. If a strictly tighter ILP formulation is found, numerical results justify the quality of the reference formulation for the problem in the Branch&Bound convergence. The quality of the continuous relaxation allows to design rounding heuristics, it offers perspectives to design matheuristics.

Keywords: Optimization · Integer Linear Programming · Linear ordering · Median of permutations · Consensus ranking · Polyhedral analysis

1 Introduction

A bridge between optimization and Machine Learning (ML) exists to optimize training parameters of ML models, using continuous optimization and meta-heuristics [18,19]. Discrete and exact optimization, especially Integer Linear Programming (ILP), is also useful to model and solve specific variants of clustering or selection problems for ML [5,6]. In this paper, another application of ILP to learning is studied: the Linear Ordering Problem (LOP). LOP is used in social choice theory to define a common consensus ranking based on pairwise preferences. If many applications deal with a small number of items to rank, bio-informatics applications solve specific LOP instances of large size as medians of permutations [2,3]. An ILP formulation is available for LOP with constraints defining facets [10,11]. An extension of LOP considering ties relies on this ILP formulation [2]. Current and recent works focus on consensus ranking for biological datasets, and use specific data characteristics of these median of permutation problems for an efficient resolution [1,14]. This paper analyzes the limits of state-of-the art ILP solvers to solve LOP instances. Several alternative ILP formulations are designed using recent results on the Asymmetric Traveling Salesman Problem (ATSP) from [15]. Comparison of Linear Programming (LP) relaxations illustrates and validates polyhedral analyses, as in [15]. The practical implication of polyhedral work is analyzed on the resolution using modern ILP solvers, as in [4]. Lastly, the quality of LP relaxation is used to design variable fixing matheuristics, as in [7]. Variants of variable definitions and ILP Formulations are recalled in Tables 1 and 2.

Table 1. Definitions of variables in the ILP formulations

Variables	Definitions
$x_{i,j} \in \{0,1\}$	$x_{i,j} = 1$ iff item $i \neq j$ is ranked before j
$y_{i,j} \in \{0,1\}$	$y_{i,j} = 1$ iff item $i \neq j$ is ranked immediately before j
$f_i \in \{0,1\}$	$f_i = 1$ iff item i is the first item of the ranking
$l_i \in \{0,1\}$	$f_i = 1$ iff item i is the last item of the ranking
$n_i \in [0, N-1]$	$n_i - 1$ gives the position of item i in the ranking
$z_{i,j,k} \in \{0,1\}$	$z_{i,j,k} = 1$ for $i \neq k$ and $j \neq k$ iff i is ranked before j and item j is ranked immediately before k
$z'_{i,j,k} \in \{0,1\}$	$z'_{i,j,k} = 1$ for $i \neq j \neq k$ iff i is ranked before j and j is before k

2   Problem Statement and Reference ILP Formulation

LOP consists in defining a permutation of N items indexed in $[1; N]$, while maximizing the likelihood with given pairwise preferences. $w_{i,j} \geqslant 0$ denotes the preference between items i and j: i is preferred to j if $w_{i,j}$ is higher than $w_{j,i}$. A ranking is evaluated with the sum of $w_{i,j}$ in the $\frac{N(N-1)}{2}$ pairwise preferences it implies. Each permutation of $[1; N]$ encodes a solution of LOP, there are $N!$ feasible solutions. The reference ILP formulation, given and analyzed in [10,11], uses binary variables $x_{i,j} \in \{0,1\}$ such that $x_{i,j} = 1$ if and only if item i is ranked before item j in the consensus permutation. With such encoding, one computes the rank of each item i with $1 + \sum_{j \neq i} x_{j,i}$. ILP formulation from [11] uses $O(N^2)$ variables and $O(N^3)$ constraints:

$$\max_{x \geqslant 0} \quad \sum_{i \neq j} w_{i,j} x_{i,j} \tag{1}$$

$$x_{i,j} + x_{j,i} = 1 \qquad \forall i < j, \tag{2}$$

$$x_{i,j} + x_{j,k} + x_{k,i} \leqslant 2 \qquad \forall i \neq j \neq k, \tag{3}$$

Constraints (2) model that either i is preferred to j, or j is preferred to i. Constraints (3) ensure that $x_{i,j}$ variables encode a permutation: if i is before j and j before k, i.e. $x_{i,j} = 1$ and $x_{j,k} = 1$, then i must be before k, i.e. $x_{i,k} = 1$ which is equivalent to $x_{k,i} = 0$ using (2). Constraints (2) and (3) are proven to be facet defining under some conditions [11].

Note that some alternative equivalent ILP were formulated. Firstly, the problem is here defined as a maximization, whereas it is considered as a minimization of disagreement in [2]. Considering $w'_{i,j} = w_{j,i}$ or $w'_{i,j} = M - w_{i,j}$, where M is an upper bound of weights $w_{i,j}$, it allows to transform minimization into maximization. Secondly, equations (3) are equivalently written as $x_{i,k} - x_{i,j} - x_{j,k} \geqslant -1$ in [2]. Formulation (3) is symmetrical, and was also used for the ATSP [17]. To see the equivalence, we use that $-x_{i,k} = x_{k,i} - 1$:

$$x_{i,k} - x_{i,j} - x_{j,k} \geqslant -1 \iff -x_{i,k} + x_{i,j} + x_{j,k} \leqslant 1$$
$$x_{i,k} - x_{i,j} - x_{j,k} \geqslant -1 \iff x_{k,i} - 1 + x_{i,j} + x_{j,k} \leqslant 1$$
$$x_{i,k} - x_{i,j} - x_{j,k} \geqslant -1 \iff x_{k,i} + x_{i,j} + x_{j,k} \leqslant 2$$

3   From ATSP to Consensus Ranking, Tighter Formulations

LOP and ATSP feasible solutions may be encoded as permutations of $[\![1; N]\!]$, order matters for cost computations. If any LOP solution is a permutation, ATSP solutions are Hamiltonian oriented cycles. LOP solutions can be projected in a cycle structure, adding a fictive node 0 such that $w_{0,i} = w_{i,0} = 0$ opening and closing the cycle: $x_{0,i} = 1$ (resp $x_{i,0} = 1$) expresses that i is the first (resp last) item of the linear ordering. This section uses polyhedral work from ATSP to tighten the reference formulation for LOP [15]. For ATSP, binary variables $y_{i,j} \in \{0, 1\}$ are defined such that $y_{i,j} = 1$ if and only if j is next item immediately after item i, for $i \neq j \in [\![1; N]\!]$. Equivalently to consider a fictive node 0, we define binary variables $f_i, l_i \in \{0, 1\}$ such that $f_i = x_{0,i} = 1$ (resp $l_i = x_{i,0} = 1$) denotes that item i is the first (resp last) in the linear ordering. Having these variables x, y, l, f induces another ILP formulation for LOP, denoted SSB for ATSP [17]:

$$\max_{x,y,f,l} \quad \sum_{i \neq j} w_{i,j} x_{i,j} \tag{4}$$

$$x_{i,j} + x_{j,k} + x_{k,i} \leqslant 2 \qquad \forall i \neq j \neq k, \tag{5}$$

$$y_{i,j} \leqslant x_{i,j} \qquad \forall i \neq j, \tag{6}$$

$$x_{i,j} + x_{j,i} = 1 \qquad \forall i < j, \tag{7}$$

$$\sum_i f_i = 1 \tag{8}$$

$$\sum_i l_i = 1 \tag{9}$$

$$l_i + \sum_{j \neq i} y_{i,j} = 1 \qquad \forall i, \tag{10}$$

$$f_i + \sum_{j \neq i} y_{j,i} = 1 \qquad \forall i, \tag{11}$$

Objective function differs from ATSP: a weighted sum of y, f, l variables is minimized for ATSP [17]. For ATSP, x variables were used only to cut subtours, whereas it is necessary for LOP to write a linear objective function. The constraints are identical for ATSP and LOP once variables x, y, f, l are used. Constraints (5) and (7) reuse (2) and (3). Constraints (10) and (11) are ATSP elementary flow constraints: for each item there is a unique predecessor and a unique successor, 0 as node successor or predecessor implies variables f_i, l_i are used. Unicity constraints (8) and (9) are ATSP elementary flow constraints arriving to and leaving from the fictive node 0. SSB can be tightened in the SSB2 formulation, replacing constraints (5) by tighter constraints (12) from [17]:

$$\forall i \neq j \neq k, \quad x_{i,j} + y_{i,j} + x_{j,k} + x_{k,i} \leqslant 2 \tag{12}$$

Sub-tours between two cities (or items) may have a crucial impact in the resolution, as in [5]. Having variables x and constraints (5) and the tighter variants implies the other sub-tours between two items. Indeed, $y_{i,j} + y_{j,i} \leqslant x_{i,j} + x_{j,i} = 1$. Constraints (13) are sub-tour cuts between node 0 and each item $i > 0$, these are known to tighten strictly SSB2 formulation [17]:

$$\forall i, \quad f_i + l_i \leqslant 1 \tag{13}$$

Another formulation was proposed for ATSP without constraints (5), but with linking constraints $y_{i,j} - x_{k,j} + x_{k,i} \leqslant 1$, to induce the same set of feasible solution [9]. These constraints can be tightened in two different ways:

$$\forall i \neq j \neq k, \quad y_{i,j} + y_{j,i} - x_{k,j} + x_{k,i} \leqslant 1 \tag{14}$$

$$\forall i \neq j \neq k, \quad y_{i,j} + y_{k,j} + y_{i,k} - x_{k,j} + x_{k,i} \leqslant 1 \tag{15}$$

Tightening only with (14) and (15) induce respectively GP2 and GP3 formulations for ATSP [9]. A strictly tighter formulation, denoted GP4, is obtained with both sets of constraints [15]. A strictly tighter formulation is also obtained adding (12) to (14) and (15) for ATSP. Numerical issues are to determine whether the quality of LP relaxation is significantly improved after tightening for LOP.

4 Other ILP Reformulations

In this section, alternative ILP reformulations for LOP are provided, adapting other formulations from ATSP. Firstly, a formulation with $O(N^2)$ variables and constraints is given, before three-index formulations with $O(N^3)$ variables.

4.1 ILP Formulation with $O(N^2)$ Variables and Constraints

Similarly with MTZ formulation [13], $O(N)$ additional variables $n_i \in [0, N-1]$ directly indicate the position of the item in the ranking :

$$\max_{x,n \geqslant 0} \quad \sum_{i \neq j} w_{i,j} x_{i,j} \tag{16}$$

$$x_{i,j} + x_{j,i} = 1 \qquad \forall i < j, \tag{17}$$

$$n_j + N \times (1 - x_{i,j}) \geqslant n_i + 1 \, \forall i \neq j, \tag{18}$$

$$n_i + \sum_{j \neq i} x_{i,j} = N - 1 \qquad \forall i, \tag{19}$$

$$n_i \in [0, N-1] \qquad \forall i \tag{20}$$

Note that as for MTZ formulation, variables n_i can be declared as continuous, feasibility of (18) and bounds (20) implies $n_i \in [\![0, N-1]\!]$. Objective function (16) and constraints (17) are unchanged. Constraints (18) are similar with MTZ constraints: if i is ranked before j, i.e. $x_{i,j} = 1$, then it implies $n_j \geqslant n_i + 1$, N is a "big M" in this linear constraint. If (17) and (18) are sufficient to induce feasible solutions for the ILP, constraints (19) complete (18) without using any "big M". Indeed, $c_i = \sum_{j \neq i} x_{i,j}$ counts the number of items after i, so that for each i, $c_i + n_i = N - 1$. As big M constraints are reputed to be weak and inducing poor LP relaxations, a numerical issue is to determine the difference with the continuous relaxation when relaxing also constraints (18) and (19).

Note that this relaxation has trivial optimal solutions, considering $x_{i,j} = 1$ and $x_{j,i} = 0$ for $i \neq j$ such that $w_{i,j} \geqslant w_{j,i}$. Hence, following upper bound is valid, and also larger than any LP relaxation for LOP:

$$UB = \sum_{i<j} \max(w_{i,j}, w_{j,i}) \tag{21}$$

4.2   Three-Index Flow Formulation

Another three index formulation, tighter than GP2, GP3 and GP4, was proposed for ATSP [9]. Adapting this formulation to LOP, one uses binary variables $z_{i,j,k} \in \{0,1\}$ for $i \neq k$ and $j \neq k$ defined with $z_{i,j,k} = 1$ if and only if i is ranked before j (not necessarily immediately before) and j is ranked immediately before k. First and last items are still marked with binaries $f_i, l_i \in \{0,1\}$. Binaries $x_{i,j}, y_{i,j} \in \{0,1\}$ are then defined by $x_{i,j} = \sum_k z_{i,j,k} + l_j$ and $y_{i,j} = z_{i,i,j}$.

$$\max_{z,l,f \geqslant 0} \quad \sum_{i \neq j} w_{i,j} \left(l_i + \sum_k z_{i,j,k} \right) \tag{22}$$

$$l_i + \sum_k z_{i,j,k} + l_j + \sum_k z_{j,i,k} = 1 \quad \forall i < j, \tag{23}$$

$$\sum_i f_i = 1 \tag{24}$$

$$\sum_i l_i = 1 \tag{25}$$

$$l_i + \sum_{j \neq i} z_{i,i,j} = 1 \quad \forall i, \tag{26}$$

$$f_i + \sum_{j \neq i} z_{j,j,i} = 1 \quad \forall i, \tag{27}$$

$$z_{i,j,k} \leqslant z_{j,j,k} \quad \forall i,j,k, \tag{28}$$

Constraints (23) and (24)–(27) are respectively constraints (7) and (8)–(11) replacing x, y occurrences by the linear expressions using z, l variables. A similar operation allows to write the objective function (22) using z, l variables. Constraints (28) model that $z_{i,j,k} = 1$ implies that j is ranked just before k and thus $z_{j,j,k} = 1$. This formulation has $O(N^3)$ variables and $O(N^3)$ constraints only because of constraints (28). It is possible to preserve the validity of the ILP while having only $O(N^2)$ constraints replacing flow constraints (28) by the aggregated version:

$$\forall j, k, \quad \sum_i z_{i,j,k} \leqslant N z_{j,j,k} \tag{29}$$

4.3   Another Three-Index Flow Formulation

In another three-index formulation, binary variables $z'_{i,j,k} \in \{0,1\}$ are defined such that for $i \neq j \neq k$, $z'_{i,j,k} = 1$ if and only if items i, j, k are ranked in this order. In this ILP formulation, we keep variables $x_{i,j}$, it induces the valid ILP formulation for LOP:

$$\max_{x, z \geqslant 0} \quad \sum_{i \neq j} w_{i,j} x_{i,j} \tag{30}$$

$$3z'_{i,j,k} \leqslant x_{i,j} + x_{j,k} + x_{i,k} \qquad \forall i \neq j \neq k \tag{31}$$

$$z'_{i,j,k} + z'_{i,k,j} + z'_{j,i,k} + z'_{j,k,i} + z'_{k,j,i} + z'_{k,i,j} = 1 \quad \forall i \neq j \neq k, \tag{32}$$

Constraints (31) are linking constraints among variables x, z: $z'_{i,j,k} = 1$ implies $x_{i,j} = x_{j,k} = x_{i,k} = 1$. Constraints (32) express that each triplet $i \neq j \neq k$ is assigned in exactly one order in a permutation, replacing constraints of type $x_{k,i} + x_{i,j} + x_{j,k} \leqslant 2$. Constraints (32) induce that this ILP formulation has also $O(N^3)$ variables and $O(N^3)$ constraints. Note that a similar constraint can be defined as cut for the previous ILP formulation, with an inequality:

$$z_{i,j,k} + z_{i,k,j} + z_{j,i,k} + z_{j,k,i} + z_{k,j,i} + z_{k,i,j} \leqslant 1 \tag{33}$$

Table 2. Summary of implemented formulations, their denomination, the sets and asymptotic number of variables and constraints

Formulation	Variables	Constraints	nbVariables	nbConstraints
LOP_ref	x	(2), (3)	$O(N^2)$	$O(N^3)$
LOP_SSB2	x, f, l, y	(6)–(11), (12, (13)	$O(N^2)$	$O(N^3)$
LOP_GP3	x, f, l, y	(6)–(11), (15)	$O(N^2)$	$O(N^3)$
LOP_MTZ	x, n	(17), (19)	$O(N^2)$	$O(N^2)$
LOP_flowGP	z, f, l	(23)–(27), (28)	$O(N^3)$	$O(N^3)$
LOP_flowGP_aggr	z, f, l	(23)–(27), (29)	$O(N^3)$	$O(N^2)$
LOP_flow2	x, z'	(31), (32)	$O(N^3)$	$O(N^3)$

5   Computational Experiments and Results

Numerical experiments were proceeded using a workstation with a dual processor Intel Xeon E5-2650 v2@2.60GHz, using at most 16 cores and 32 threads in total. Cplex version 20.1 was used to solve LPs and ILPs. Cplex was called using OPL modeling and OPL script languages. LocalSolver in its version 10.5 was used as a heuristic solver benchmark to compare primal solutions when optimal solutions are not proven. The maximal time limit for Cplex and LocalSolver was set to one hour, Cplex was used with its default parameters. For reuse and reproducibility, OPL and LocalSolver codes and generated instances are available online at https://github.com/ndupin/linearOrdering.

5.1   Data Generation and Characteristics

It was necessary to generate specific instances for this study. As mentioned by [1,14], instance characteristics are crucial in the resolution difficulty. In many social choice applications and datasets, N is small, exact resolution with formulation LOP_ref is almost instantaneous. For the biological application, N is very large but median of permutations among similar permutations is easier than general instances. In the extreme case where $w_{i,j}$ coefficients encode a permutation (median of 1-permutation, trivial problem), trivial bounds UB give the optimal value, and LP relaxations of every ILP formulation give the integer optimal solution. For this numerical study, as in [15], quality of polyhedral descriptions are analyzed on the implications on the quality of LP relaxation using diversified directions of the objective function. Three generators were used for this study:

- `aleaUniform` (denoted aUnif): $w_{i,j}$ for $i \neq j$ are randomly generated with a uniform law in $[\![0, 100]\!]$.
- `aleaSum100` (denoted aSum): uniform generation in $[\![0, 100]\!]$ such that $w_{i,j} + w_{j,i} = 100$: for $i < j$ $w_{i,j}$ is randomly generated in $[\![0, 100]\!]$ and $w_{j,i}$ is then set to $w_{j,i} = 100 - w_{i,j}$.
- `aleaShuffle` (denoted aShuf): $\max(N/2, 20)$ random permutations are generated (with Python function shuffle), $w_{i,j}$ are then computed using Kendall-τ distance and Kemeny ranking, as in [1–3].

A fourth generator was coded, as in `aleaShuffle`, but generating small perturbations around a random permutation. Actually, the results were very similar for ILP formulations to the 1-median trivial instances. Real-life structured instances for median of permutations are much easier than random instances. The generators allow to analyze the impact of structured instances.

Number of items N was generated with values $N \in \{20, 30, 40, 50, 100\}$. For $N \in \{20, 30, 40\}$, the Best Known Solution (BKS) are optimal solutions proven by Cplex. For $N \in \{50, 100\}$, LocalSolver always provides the BKS. There is also no counter-example where LocalSolver does not find a proven optimal solution in one hour, we note that LocalSolver is also efficient in short time limits. For each generator and value of N, 30 instances are generated and results are given in average for each group of 30 similar instances, with the denomination XX$-N$ where XX $\in$ {aUnif,aSum,aShuf}. Lower and upper bounds $v(i)$ on instance i are compared with gaps to BKS, denoted BKS(i):

$$gap = \frac{|\, v(i) - BKS(i)\,|}{BKS(i)} \qquad (34)$$

5.2   Comparing LP Relaxations

To analyze the quality of polyhedral descriptions recalled in Table 2, Table 3 presents gaps of LP relaxations of ILP formulations for LOP and the naive upper bound (21). Table 4 presents the computation time for LP relaxations,

Table 3. Comparison of the average gaps to the BKS for the LP relaxations of formulations recalled in Table 2 and the naive upper bound (21)

Instances	(21)	ref/SSB2	GP3	MTZ	flow-GP	flow-GP-agg	flow2
aUnif-20	12,86%	0,02%	10,49%	10,84%	5,22%	11,53%	12,86%
aUnif-30	14,86%	0,17%	13,20%	13,39%	7,34%	13,98%	14,86%
aUnif-40	16,98%	0,60%	15,68%	15,78%	9,22%	16,16%	16,98%
aUnif-50	17,84%	1,12%	16,80%	16,86%	10,22%	17,17%	17,84%
aUnif-100	21,65%	3,17%	21,10%	21,11%	–	21,25%	21,65%
aSum-20	19,00%	0,05%	15,56%	16,13%	7,26%	17,23%	19,00%
aSum-30	21,96%	0,31%	19,53%	19,84%	10,25%	20,45%	21,96%
aSum-40	24,40%	1,18%	22,52%	22,70%	12,53%	23,21%	24,40%
aSum-50	26,24%	2,25%	24,71%	24,83%	14,16%	25,23%	26,24%
aSum-100	31,44%	4,97%	30,64%	30,65%	–	30,84%	31,44%
aShuf-30	1,93%	0,00%	1,39%	1,42%	0,29%	1,89%	1,93%
aShuf-40	1,44%	0,00%	1,18%	1,18%	0,42%	1,42%	1,44%
aShuf-50	1,41%	0,00%	1,18%	1,18%	0,44%	1,40%	1,41%
aShuf-100	1,80%	0,02%	1,65%	1,55%	–	1,80%	1,80%

to highlight the impact of the number of variables and constraints recalled in Table 2. These tables illustrate the difficulty of instances, aShuf are easy instances with good naive upper bounds and LP relaxations. Datasets aSum and aUnif induce more difficulties with worse continuous bounds, and aSum is even more difficult than aUnif.

Contrary to ATSP where GP2, GP3, SSB2 are not redundant [15], (3) induces much better LP relaxations for LOP than (14) and (15). Adding (14) and (15) in ILP formulations with (3) or (12) does not induce any difference in the quality of LP relaxation. It explains why in Table 2, we remove constraints of type (3) to compare quality of LP relaxations. An explanation is the different nature of LOP and ATSP problems because of different objective functions: if polyhedrons defined by constraints are identical, objective functions with weighted sums in x or y change the projection on the space of interest.

Flow formulation flow-GP improves significantly the quality of LP relaxation of GP3, as for the ATSP, but it is still significantly worse than SSB formulations. Computation time of LP relaxation is much higher with flow-GP, computations were stopped in one hour without termination for $N = 100$. With aggregation (29) instead of (28), LP relaxation is computed quickly, but the quality of LP relaxation is dramatically decreased, the continuous bounds are close to the naive upper bounds (21). MTZ adaptation has the quickest LP relaxation, but the continuous bounds are close to the ones of GP3. Last flow formulation always provides exactly the naive upper bounds (21), constraints (33) do not tighten flow-GP formulation, this result differ from [16].

Table 4. Comparison of the average time (in seconds) to compute LP relaxations for ILP formulations recalled in Table 2

Instances	ref	SSB2	GP3	MTZ	flow-GP	flow-GP-agg	flow2
aUnif-20	0,04	0,14	0,32	0,00	1,21	0,06	0,07
aUnif-30	0,28	0,75	1,08	0,01	7,62	0,18	0,25
aUnif-40	0,49	2,27	3,58	0,03	38,60	0,53	0,83
aUnif-50	0,95	5,59	10,55	0,12	173,49	1,36	2,32
aUnif-100	26,63	278	839	1,04	–	18,24	54,61
aSum-20	0,04	0,17	0,33	0,00	1,20	0,06	0,07
aSum-30	0,29	0,77	1,08	0,01	7,48	0,19	0,25
aSum-40	0,50	2,12	3,43	0,03	37,10	0,55	0,83
aSum-50	0,95	5,65	10,74	0,12	168,24	1,40	2,27
aSum-100	27	282,46	819	1,75	–	17,40	55,63
aShuf-30	0,06	0,24	1,04	0,01	6,18	0,18	0,27
aShuf-40	0,16	0,84	3,73	0,04	26,86	0,49	0,74
aShuf-50	0,33	2,17	11,63	0,13	98,88	1,32	2,13
aShuf-100	17,7	353,5	1360	1,76	–	16,45	51,34

LP relaxation of LOP_ref is of an excellent quality, which illustrates polyhedral results and proven facets from [11]. In Table 2, LOP_ref and SSB2 formulations have the same values: except on three instances, LP relaxation are the same (with a tolerance to numerical errors on the last digit). On instance number 27 in aUnif-20 and instances number 17 and 29 in aSum-20, SSB2 improves the reference formulation around 0.01%, making a difference of one unit in the integer ceil rounding of the continuous relaxation. With additional experiments, the difference is only due to (3) instead of (12), no difference was observe adding only (13). These results proves that LOP_SSB2 is in theory strictly tighter than LOP_ref, but with small and rare improvements.

5.3   Comparing Branch and Bound Convergences

Now, we compare the impact of modeling LOP with LOP_ref and LOP_SSB2, in the Branch&Bound (B&B) convergence. Table 5 analyzes the impact of Cplex cuts and heuristics at the root node, before branching in the B&B tree. If LOP_SSB2 improves slightly LP relaxation quality, the open question is to determine if additional variables and constraints help modern ILP solvers detecting other structures for cut generation, as in [4]. For LOP, computations at the root node of B&B tree are much slower with SSB2, coherently with the higher number of variables, but the efficiency of cuts and primal heuristics is significantly worse with the heavier SSB2 formulation. Having a larger ILP model, slower matrix operations for generation of cutting planes are needed by Cplex, and this stops earlier cuts that would have been generated using LOP_ref formulation, the size

Table 5. Comparison of Lower Bounds (LB) and Upper Bounds (UB) of formulations LOP_ref and LOP_SSB2 after Cplex cuts and heuristics at the root node (i.e. before branching). Common UB with the LP relaxation are also provided for comparison.

	LP ref,SSB2	UB	LB ref	time	UB	LB SSB2	time
aUnif-20	0,02%	0,00%	0,00%	0,1	0,00%	0,00%	0,4
aUnif-30	0,17%	0,00%	0,00%	1,5	0,09%	0,50%	14
aUnif-40	0,60%	0,44%	0,22%	30,5	0,52%	4,58%	159
aUnif-50	1,12%	0,96%	0,82%	212,9	0,98%	5,27%	1336
aUnif-100	3,17%	3,07%	5,49%	3600	3,15%	7,37%	3600
aSum-20	0,05%	0,00%	0,00%	0,12	0,00%	0,00%	0,55
aSum-30	0,31%	0,02%	0,02%	2,0	0,16%	0,79%	20
aSum-40	1,18%	0,75%	0,32%	64	0,95%	5,08%	296
aSum-50	2,25%	1,82%	0,95%	297	1,95%	6,72%	989
aSum-100	4,97%	4,82%	6,87%	3600	4,95%	9,62%	3600
aShuf-30	0,00%	0,00%	0,00%	0,14	0,00%	0,00%	0,86
aShuf-40	0,00%	0,00%	0,00%	0,37	0,00%	0,00%	2,9
aShuf-50	0,00%	0,00%	0,00%	0,90	0,00%	0,00%	8
aShuf-100	0,02%	0,02%	0,02%	477	0,02%	6,02%	3395

of ILP matrix is crucial here, contrary to [4]. Table 5 shows that few improvement of LP relaxation is provided at the root node of B&B tree, cuts are not very efficient to improve the LP relaxation, which was of a good quality. This explains the difference in the B&B convergence in one hour allowing branching, LOP_ref formulation is largely superior. For some instances with $N = 40$ or $N = 50$, LOP_ref can converge in ten minutes whereas a significant gap between lower and upper bounds remains after one hour for LOP_SSB2. This definitively validates the LOP_ref formulation as baseline ILP model for [2].

5.4   Variable Fixing Heuristics

The excellent quality of the LP relaxation with LOP_ref formulation allows to use continuous solutions of LP relaxation to design primal heuristics as in [7]. Variable Fixing (VF) denotes here a heuristic reduction of the search space based on the LP relaxation, to set integer values to variables in the ILP resolution. One may use a VF preprocessing for variables with an integer value in the continuous relaxation, expecting that these integer decisions are good. Generally, it makes a difference to apply VF preprocessing on zeros and ones in the LP relaxation, as in [7]. There are in general many possibilities of VF preprocessing, considering also specific rules to select a subset of variable to fix [7].

For LOP, imposing $x_{i,j} = 1$ (resp 0) implies $x_{j,i} = 0$ (resp 1) with constraints (2), so that fixing only ones or only zeros are equivalent to fix all the integer variables, contrary to [7]. Note that constraints (3) may induce continuous solutions $x_{i,j} = x_{j,k} = x_{k,i} = 2/3$, so that rounding such variables induces infeasibility on (3) constraints. This property does not hold when rounding to ones only variables that are superior to 0.7 Hence, two VF strategies were implemented, on one hand fixing the integer value, and on the other hand considering the threshold for rounding to 0.8. Actually, there were slight differences for these two strategies. Experiments also used the quick MTZ relaxation, this significantly degraded the performance of the VF heuristic.

Table 6 compares the gap to BKS and computation time using the VF preprocessing to LOP_ref formulation. For small and easy instances where LOP_ref gives optimal solutions, the degradation of the objective function is small with the VF heuristic, speeding up significantly the computation time. For the largest instances with $N = 100$, VF matheuristic is significantly better than the exact resolution, illustrating the difficulty of the ILP solver to find good primal solutions with its primal heuristics. The primal solutions of matheuristic are in this case also significantly worse than the ones of LocalSolver, the VF speed up is not sufficient to reach an advanced phase of the B&B convergence.

Table 6. Comparison of gaps to BKS and computation time of Cplex in ILP solving using the reference formulation, without and with Variable Fixing (VF) preprocessing on integer values in the LP relaxation of the reference formulation. BKS are optimums for $N \leqslant 40$, for $N \geqslant 50$ BKS were given by LocalSolver

Instances	LB	time (sec)	LB	time (sec)
	ref		ref + VF	
aUnif-20	0,00%	0,1	0,01%	0,04
aUnif-30	0,00%	1,5	0,04%	0,62
aUnif-40	0,00%	30,5	0,17%	13
aUnif-100	5,49%	3600	2,36%	3600
aSum-20	0,00%	0,13	0,05%	0,06
aSum-30	0,00%	2,1	0,09%	0,77
aSum-40	0,00%	63,5	0,43%	12,7
aSum-100	6,87%	3600	3,14%	3600
aShuf-30	0,00%	0,13	0,00%	0,03
aShuf-40	0,00%	0,37	0,00%	0,08
aShuf-50	0,00%	0,92	0,00%	0,12
aShuf-100	0,00%	1820	0,00%	35,7

6   Conclusions and Perspectives

If the reference ILP formulation seemed to be improvable using ATSP results, only a slightly tighter ILP formulation is obtained after this reformulation work.

Analyzing the ILP convergence with a modern ILP solver shows that the LP relaxation is of an excellent quality with the reference formulation, but is fewly improved after with cuts and branching. Note that these reformulation issues were an open question raised by [12]. Also, primal heuristics are not efficient on the problem, a basic VF matheuristic significantly improves the primal solutions for difficult instances. Furthermore, this paper illustrates the graduated difficulty of instances, structured instances from the biological application as median of permutations are easier that random instances of LOP.

These results offer perspectives for the biological application also with the extension with ties [1]. Matheuristics can be used in this context, combined to specific reduction space operators related to the easier median of permutation instances [1,14]. Perspectives are also to combine matheuristics and local search approaches which are efficient for the problem, as shown by LocalSolver benchmark on this study, and also by [8,12], to solve larger instances ($N \geqslant 100$).

References

1. Andrieu, P., et al.: Efficient, robust and effective rank aggregation for massive biological datasets. Future Gener. Comput. Syst. **124**, 406–421 (2021)
2. Brancotte, B., Yang, B., Blin, G., Cohen-Boulakia, S., Denise, A., Hamel, S.: Rank aggregation with ties: experiments and analysis. Proc. VLDB Endowment (PVLDB) **8**(11), 1202–1213 (2015)
3. Cohen-Boulakia, S., Denise, A., Hamel, S.: Using medians to generate consensus rankings for biological data. In: Bayard Cushing, J., French, J., Bowers, S. (eds.) SSDBM 2011. LNCS, vol. 6809, pp. 73–90. Springer, Heidelberg (2011). https://doi.org/10.1007/978-3-642-22351-8_5
4. Dupin, N.: Tighter MIP formulations for the discretised unit commitment problem with min-stop ramping constraints. EURO J. Comput. Optim. **5**(1), 149–176 (2017)
5. Dupin, N., Parize, R., Talbi, E.: Matheuristics and column generation for a basic technician routing problem. Algorithms **14**(11), 313 (2021)
6. Dupin, N., Talbi, E.: Machine learning-guided dual heuristics and new lower bounds for the refueling and maintenance planning problem of nuclear power plants. Algorithms **13**(8), 185 (2020)
7. Dupin, N., Talbi, E.: Parallel matheuristics for the discrete unit commitment problem with min-stop ramping constraints. Int. Trans. Oper. Res. **27**(1), 219–244 (2020)
8. Garcia, C., Pérez-Brito, D., Campos, V., Martí, R.: Variable neighborhood search for the linear ordering problem. Comput. OR **33**(12), 3549–3565 (2006)
9. Gouveia, L., Pires, J.: The asymmetric travelling salesman problem and a reformulation of the Miller-Tucker-Zemlin constraints. Euro. J. Oper. Res. **112**(1), 134–146 (1999)
10. Grötschel, M., Jünger, M., Reinelt, G.: A cutting plane algorithm for the linear ordering problem. Oper. Res. **32**(6), 1195–1220 (1984)
11. Grötschel, M., Jünger, M., Reinelt, G.: Facets of the linear ordering polytope. Math. Program. **33**(1), 43–60 (1985). https://doi.org/10.1007/BF01582010
12. Martí, R., Reinelt, G.: Exact and Heuristic Methods in Combinatorial Optimization: A Study on the Linear Ordering and the Maximum Diversity Problem. Springer, Cham (2022)

13. Miller, C., Tucker, A., Zemlin, R.: Integer programming formulation of traveling salesman problems. J. ACM **7**(4), 326–329 (1960)
14. Milosz, R., Hamel, S.: Space reduction constraints for the median of permutations problem. Discrete Appl. Math. **280**, 201–213 (2020)
15. Öncan, T., Altınel, I., Laporte, G.: A comparative analysis of several asymmetric traveling salesman problem formulations. Comput. OR **36**(3), 637–654 (2009)
16. Peschiera, F., Dell, R., Royset, J., Haït, A., Dupin, N., Battaïa, O.: A novel solution approach with ML-based pseudo-cuts for the flight and maintenance planning problem. OR Spectrum **43**(3), 635–664 (2021)
17. Sarin, S., Sherali, H., Bhootra, A.: New tighter polynomial length formulations for the asymmetric traveling salesman problem with and without precedence constraints. Oper. Res. Lett. **33**(1), 62–70 (2005)
18. Talbi, E.: Combining metaheuristics with mathematical programming, constraint programming and machine learning. Ann. OR **240**(1), 171–215 (2016)
19. Talbi, E.: Machine learning into metaheuristics: a survey and taxonomy. ACM Comput. Surv. (CSUR) **54**(6), 1–32 (2021)

SHAMan: A Versatile Auto-tuning Framework for Costly and Noisy HPC Systems

S. Robert[1], S. Zertal[2(✉)], and P. Couvée[1]

[1] Atos BDS R&D Data Management, Echirolles, France
{sophie.robert,philippe.couvee}@atos.net
[2] University of UPSacaly-UVSQ, Guyancourt, France
soraya.zertal@uvsq.fr

1 Introduction

Most of the software of modern computer systems come with many configurable parameters that control the system's behavior and its interaction with the underlying hardware. These parameters are challenging to tune by solely relying on field insight and user expertise, due to huge spaces and complex, non-linear system behavior. Besides, the optimal configuration often depends on the current workload, and parameters must be changed at each environment variations. Consequently, users often have to rely on the default parameters given by the provider, and do not take advantage of the possible performance with a more appropriate parametrization of their tuned system. The more complex these systems are, the more important the tuning becomes, as the components interact with each other in ways that are hard to grasp by the human mind. As architecture becomes more and more service oriented, the number of components per system increases exponentially, along with the number of tunable parameters. This problem is particularly observed within HPC systems with hundreds of devices assembled to build very complex and highly configurable supercomputers. As performance is the major concern in this field, each component must be finely tuned, which is almost impossible to achieve solely through field expertise. An additional constraint is the often noisy setting, because making resources exclusive is expensive and goes against the current highly shared programming environments. Automatic tuning methods thus must need to take into account this possible interference on the tuned application, which degrades the performance of classical auto-tuning heuristics. Faced with the inability of relying solely on users to take adequate decisions for the parametrization of complex computer systems, new tuning methods have emerged from various computer science communities to automate parameter selection depending on the current workload. Because they do not require any human intervention, these approaches are commonly called *auto-tuning* methods, a term which encompasses a broad range of methods related to the optimization and machine learning field. Throughout the years, they have been successfully applied to a wide range of systems, such as storage systems, database management systems and compilers.

B. Dorronsoro et al. (Eds.): OLA 2022, CCIS 1684, pp. 87–104, 2022.
https://doi.org/10.1007/978-3-031-22039-5_8

In this paper, we introduce a new Open Source software, called SHAMan (**S**mart **HPC** **A**pplication **Man**ager), which provides an out-of-the-box Web application to perform black-box auto-tuning of custom computer components running on a distributed system, for an application submitted by the user. The framework integrates three state-of-the-art black-box optimization heuristics, as well as resampling-based noise reduction strategies to deal with the interference of shared resources, and pruning strategies to limit the time spent by the optimization process. It is to our knowledge the only generalistic optimization framework specifically tailored to find the optimum parameters of configurable HPC systems, by taking into account their specific constraints.

This paper is organized as follows. We introduce in Sect. 2 the works related to ours, and discuss the improvements provided by SHAMan compared to the state-of-the-art. Section 3 introduces the theoretical context of black-box optimization for noisy and expensive systems. In Sect. 4 we present the different features of SHAMan and its software architecture. In Sect. 5, we present the advantages of using SHAMan on three use-cases by tuning two different I/O accelerators and OpenMPI collectives. Finally, Sect. 6 concludes the paper and gives insights into planned further works.

2   Related Works and Software

Within the HPC community, auto-tuning has gained a lot of attention for tuning particular HPC application and improve their portability across architectures [14]. Seymour et al. [35] and Knijnenburg et al. [23] provide a comparison of several random-based heuristic searches (Simulated annealing, genetic algorithms ...) that have provided some good results when used for code auto-tuning. Menon et al. [26] use Bayesian Optimization and suggest the framework HiPerBOT to tune application parameters as well as compiler runtime settings. HPC systems energy consumption can also benefit from Bayesian Optimization, as Miyazaki et al. have shown in [28] that an auto-tuner based on a combination of Gaussian Process regression and the Expected Improvement acquisition function has raised their cluster to the Green500 list. The MPI community has also shown the superiority of a hill-climbing black-box algorithm over an exhaustive sampling of the parametric space in [20] and [41].

In terms of auto-tuning frameworks, several have been proposed recently in different domain where optimization is required. The Machine Learning community has proposed several frameworks to find the parameters that return the best prediction scores for a given model and dataset. Among the most popular frameworks, we can cite Optuna [12], which relies on Tree Parzen Estimators to perform the optimization. Autotune [24] is an other framework which supports several black-box optimization techniques. Scikit-Optimize [9] which supports a wide range of surrogate modeling techniques. The SHERPA [21] library provides different optimization algorithms with the possibility to add new ones. It also comes with a back-end database and a small Web Interface for experiment visualization. GPyOpt [13] is another library for users who wish to use Bayesian

Optimization. Finally, the framework TPOT [25] relies on genetic algorithms for the optimization of Machine Learning pipelines. Within the MPI community, the two most famous commercial implementations come with their own tuning tool: OPTO [15] is the standard tool used by the Open MPI community for tuning MCA parameters, and similarly mpitune [6] from Intel MPI. These methods only include exhaustive grid search, making these tools slow to use for tuning expensive HPC applications. The main drawbacks identified with these already existing frameworks concern the difficulty of integrating these libraries for purpose other than the ones they were designed for by using them for HPC tuning. It is also difficult to enhance them with other optimization techniques, such as noise reduction strategies. In addition, none of them offer a satisfying Web Interface allowing an easy manipulation of the software.

Faced with the highlighted deficiencies of the existing solutions, we have developed our own framework, and provide these main contributions and features:

(a) **Versatility:** it can handle a wide range of use-cases, and new components can be registered through a generalist configuration file.
(b) **Accessibility:** the optimization engine is accessible through a Web Interface.
(c) **Optimization diversity:** as different heuristics work differently for different systems, our framework provides several state-of-the-art heuristics.
(d) **Easy extention:** the optimization engine uses a plug-in architecture and the development of the heuristic is thus the only development cost.
(e) **Integration within the hpc ecosystem:** the framework relies on the Slurm workload manager [22] to run hpc applications. The microservice architecture enables it to have no concurrent interactions with the cluster and the application itself. It is not intrusive allowing users to launch applications on their own. Also, no privileged rights are required to use the software.
(f) **Customizable target measure:** the optimized target function can be defined on a case-per-case basis to allow the optimization of various metrics.
(g) **Integration of noise reduction strategies:** because of the highly dynamic nature and the complexity of applications and software stacks, running in parallel on shared resources are subject to many interference, which results in a different performance measure for each run even with the same system's parametrization. Noise reduction strategies are included in the framework to perform well even in the case of strong interference.
(h) **Integration of pruning strategies:** runs with unsuited parametrization are aborted, to speed-up the convergence process.

3   Theoretical Background

SHAMan relies on black-box optimization, which consists in treating the tuned system as a black-box, deriving insight only from the relationship between the input and the output parameters, as described by the optimization loop in Fig. 1.

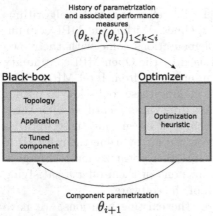

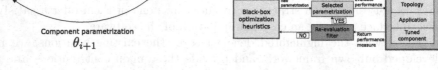

Fig. 1. Schematic representation of the optimization loop, without

Fig. 2. Schematic representation of the optimization loop, with noise reduction through resampling

3.1 An Overview of the Optimization Loop

Black-box optimization refers to the optimization of a function f with unknown properties in a minimum of evaluations, without making any assumption on the function. The only available information is the history of the black-box function, which consists in the previously evaluated parameters and their corresponding objective value. Given a budget of n iterations, the problem can be transcribed as:

$$Find\{p_i\}_{1 \leq i \leq n} \in \mathbb{P} \; s.t. \; | \; min(f(p_i)_{1 \leq i \leq n}) - min(f) \leq \epsilon \; | \tag{1}$$

- f the function to optimize
- $\mathbb{P}$ the parameter space
- ϵ a convergence criterion between the found and the estimated minimum

Every black-box optimization process starts with the selection of the initial parameters for the algorithm. An acceptable initialization starting plan should respect two properties [19,40]: the space's constraints and the non-collapsible property. The space constraints are shaped by the possible values that can be taken by the parameters. The non-collapsible property specifies that no parametrization can have the same value on any dimension. A design plan respecting this constraint is called a *Latin Hypercube Design* (LHD) [19]. The next step consists in a feedback loop, which iteratively selects a parametrization, evaluates the black-box function at this point and selects accordingly the next data point to evaluate. The higher procedure for searching an optimal solution in a parametric space is called an optimization *heuristic*. There are many black-box heuristics, and we have implemented the following set in our optimization engine because of their simplicity of implementation and their proven efficiency for HPC systems' tuning. A detailed motivation and description of our implementation can be found in [31] and [32].

- **Surrogate models:** Surrogate modeling consists in using a regression or interpolation technique over the currently known surface to build a computationally cheap approximation of the black-box function and to then select the most promising data point in terms of performance on this surrogate function by using an acquisition function.
- **Simulated annealing:** the simulated annealing heuristic is a hill-climbing algorithm which can probabilistically accept a solution worse than the current one.
- **Genetic algorithms:** Genetic algorithms consist in selecting a subset of parameters, among the already tested parametrizations, considering the objective value of each parametrization, and then combining them to create a new parametrization.

3.2 Stop Criteria

When running the optimization algorithm, the easiest stop criterion is based either on a budget of possible steps (exhaustion based) or on a time-out based on the maximum elapsed time for the algorithm. Once the iteration budget has been spent, the algorithm stops and returns the best found parametrization. However, while very simple to implement, this criterion can be inefficient, as it has no adaptive quality on the tuned system. Using other stop criterion to speed-up the convergence process, while still providing a good solution is thus essential for tuning expensive systems, and SHAMan integrates two different stop criteria:

Exhaustion Based Criteria. Exhaustion based criteria are criteria based on the number of allowed iterations performed by the heuristic. Once all of the possible iterations have been tried by the algorithm, the algorithm stops and the parametrization which returned the best corresponding performance measure is kept. They are the most popular in the black-box optimization literature [43] because of their simplicity of implementation and the control they give over the optimization process. However, they can be a waste of resource because they can either:

- Stop the algorithm while the maximum optimization potential has not been reached, thus not finding the optimal parametrization. The algorithm either has to be started from scratch or can resume, depending on the practical auto-tuning implementation.
- Keep the algorithm running even though the maximum potential of optimization has already been reached. This is a waste of time and resources, as the algorithm runs aimlessly without providing any improvement.

Exhaustion-based criteria thus do not provide much flexibility in the optimization process and do not have any adaptive qualities to the behavior of the system. Because of this, SHAMan integrates two other criteria based either on the value of the performance function or the value of the tested parameters:

Improvement Based Criteria. They consist in stopping the optimization process if it does not bring any improvement over a given number of iterations. Depending on the target behavior, the improvement can either be measured globally as the average of the evaluated values or locally as the change in optimum values.

- **Best improvement:** Improvement of the best objective function value is below a threshold t for a number of iterations g.
- **Average improvement:** Improvement of the average objective function value is below a threshold t for a number of iterations g.
- **Median improvement:** Improvement of the median objective function value is below a threshold t for a number of iterations g.

Movement Based Criteria. Movement based criteria consider the movement of the parametric grid as a criteria to stop the optimization. Two variations of the criteria are available in our framework:

- **Count based:** The optimization algorithm is stopped once there is less than t different parametrization evaluated over a number of iterations g.
- **Distance based:** The optimization algorithm is stopped once the distance between each parametrization goes below a certain threshold t for a number of iterations g.

3.3 Resampling for Noisy Systems

Resampling consists in adding a "resampling filter" by using a set logical rule to select which parametrization to reevaluate. A detailed schematic representation of the integration of resampling within the black-box optimization tuning loop is available in Fig. 2. The general goal of resampling is to reduce the standard deviation of the mean of an objective value in order to augment the knowledge of the impact of the parameter on the performance.

Algorithmically, we define a resampling filter as a function $\mathcal{RF}$ which takes as input an optimization's trajectory already evaluated fitness and corresponding parameters $(\theta_i \in \Theta, F(\theta_i) \in \mathbb{R})_k$, for the optimization trajectory at step k, and outputs a boolean on whether or not the last parametrization should be re-evaluated. This filter can be integrated for both initialization draws and exploitation draws, or only for exploitation ones. We make the latter choice, as we want to keep the initialization draw to test as many parametrization as possible, and if needed, let the algorithm come back to these parametrization for further investigation. Resampling is a trade-off between having a better knowledge of the space and waste some computing times on re-evaluation. We present here two of the most popular resampling algorithms in order to efficiently reevaluate a parametrization and a more exhaustive description is proposed in [36]. Three noise strategies are available in the framework:

- **Static resampling:** re-evaluates each parametrization for a fixed number of iterations.
- **Standard Error Dynamic Resampling** [18]: re-evaluates the current parametrization until the standard error of the mean falls below a set threshold.
- **An improved noise reduction algorithm:** a complex decision algorithm for re-evaluating the current parametrization [32].

3.4   Pruning of Expensive Systems

Pruning strategies consist in stopping some runs early because their parametrization is unpromising, compared to already tested parameters. Two pruning strategies are available in the framework:

- **Default based:** It consists in stopping every run that takes longer than the execution time corresponding to the default parametrization.
- **Estimator based:** It consists in stopping every run that takes longer than the value of an estimator computed on previous runs. For example, if the selected estimator is the median, the current run is stopped if its elapsed time takes longer than 50% of the already tested parametrization. This pruning only applies to runs performed after the initialization ones.

4   Software Architecture and Features

SHAMan (Smart HPC Applications Manager) performs the auto-tuning loop by parametrizing the component, submitting the job through the Slurm workload manager, and getting the corresponding execution time Using the combination of the history (parametrization and execution time) to select the next most appropriate parametrization until the stop criterion is reached.

4.1   Terminology

Throughout this section, we will use the following terms:

- **Component:** the configurable component which optimum parameters must be found.
- **Target value:** the measurement that needs to be optimized.
- **Parametric grid:** the possible parametrization defined as (minimum, maximum, step value).
- **Application:** a program that can be run on the clusters' nodes through Slurm to be tuned.
- **Budget:** the maximum number of evaluations to find the optimum value.
- **Experiment:** A combination of a component, a target value, an application and a parametrized black-box optimization algorithm that will output the best parametrization for the application and the component.

4.2   Optimization and Vizualization Procedure

The main features of SHAMan are the possibility to:

Declare a New Configurable Component and Register it for Later Optimization.
Running the command `shaman-install` with a YAML file describing a component registers it to the application and makes it possible to be optimized. This file must describe how the component is launched and declares its different parameters and how they must be used to parametrize it. After the installation process, the components are available for optimization in the launch menu.

Design and Launch an Experiment Through a Web Interface or Through a Command Line Interface. The main way is to launch the experiment through the Web interface via the menu. The user has to configure the black-box by:

1. Writing an application according to Slurm sbatch format.
2. Selecting the component and the parametric grid through the radio buttons.
3. Configuring the optimization heuristic, chosen freely among available ones. Resampling parametrization, stop criterion and pruning strategies can also be activated.
4. Selecting a maximum number of iterations and the name of the experiment.

The optimization process will begin to run and its information will be available in the exploring section of the Web application.

Another way to use SHAMan is to use it directly through a command line interface. It allows more flexibility of the different features and requires the same information as the Web interface.

Visualize Data and Results of Finished or Running Experiments. After the submission, the evolution of the running experiments can be visualized in real-time. The optimization trajectory is available through a display of the different tested parameters and the corresponding execution time, as well as the improvement brought by the best parametrization. The other metadata of the experiment are also available through side menus. Figure 3 and 4 show the tunes performance without any aggregation, then with it if the noise reduction is enabled.

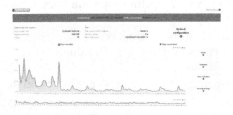   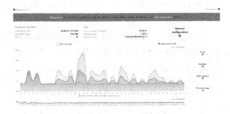

Fig. 3. Visualization of an optimization trajectory

Fig. 4. Visualization of an optimization trajectory when noise reduction is enabled

4.3   Software Architecture

The architecture of SHAMan relies on microservices, as can be seen in Fig. 5 and detailed in [33]. It is composed of several services, which can each be deployed independently:

- An optimization engine which performs the optimization tasks.
- A front-end Web application
- A back-end storage database
- A rest api enabling communications of all the services

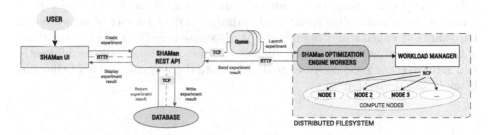

Fig. 5. General architecture of the tuning framework

4.4   Implementation Choices

SHAMan uses *Nuxt.js* as a frontend framework. The optimization engine is written in Python. The database relies on the NoSQL database management system *MongoDB*. The message broker system uses Redis [17] as a queuing system, manipulated with the ARQ Python library [1]. The API is developed in Python, using the FastAPI framework. The framework is fully tested, can be fully deployed as a stack of Docker containers [27]. The code is available on Github [10] and thoroughly documented [3].

5   Use-Cases and Results

In some of our previous works, we have shown the efficiency of SHAMan on two different use-cases belonging to I/O accelerators: a smart prefetching strategy and a burst buffer [33, 34]. To further prove the versatility of SHAMan, we tackle another difficult to tune HPC component: MPI collectives. We begin this section by summarizing the main results of our previous experiments, and then introduce the new results provided by our experiment on MPI collectives.

5.1   I/O Accelerators

I/O accelerators are software or hardware components which aim is to reduce the increasing performance gap between compute nodes and storage nodes, which can slow down I/O intensive applications. Indeed, on large supercomputers, the many compute nodes performing reads or writes can stress the storage bay and make the application wait while it performs its I/O, generating *I/O bottlenecks*. This is especially true for HPC applications that periodically save their current state (by performing *checkpoints*) which causes many writes during a short timeframe. The link between the compute and the storage node can become saturated, which slows down the application. To mitigate these problems, several I/O accelerators have been developed over the years, and we focused specifically with SHAMan on tuning two commercial implementations of I/O accelerators: a pure software one called smart read optimizer [11] and a mix of software and hardware one called smart burst buffer [2]. Because these I/O accelerators come with many parameters, they are difficult to tune, and operate very differently which make them good use-cases to demonstrate the versatility of SHAMan and black-box optimization.

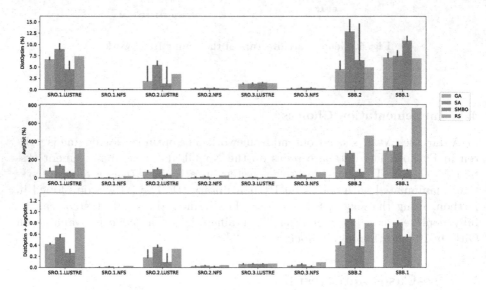

Fig. 6. Best values for DistOptim, AvgDist and the sum of both for every heuristic

For both I/O accelerators, we performed a comparative study of the impact of each black-box optimization heuristic with SHAMan and showed that surrogate models offer the best trade-off between optimization quality and stability of the trajectory, and outperforms by far a random sampler.

As displayed in Fig. 6, we show that with a distance to the true minimum inferior to 4% for every application, our auto-tuner exhibits good convergence properties. As we have found convergence rate inferior to 40 steps for reaching 5% of the optimal value, we have also demonstrated that the auto-tuner can operate in a sparse production environment for expensive systems.

5.2   Tuning MPI Collectives

MPI is a standardized and portable message-passing standard designed to function on parallel computing architectures through many implementations as Open MPI [7] and MPICH [5]. Its collective operations provide a standardized interface for performing data movements within groups of processes and come with several tunable parameters. The optimal configuration greatly depends on the size of the transmitted message [30], as well as the architecture and the topology of the target platform [42], and the default parametrization is not adapted for a wide range of cases. In particular, the Open MPI [7] implementation features a modular architecture and the selection of modules along with their parametrization is achieved through MCA parameters, which can be provided using either configuration files, command-line arguments or environment variables. This implementation is the one we will be focusing on, by considering the subset of parameters related to the `coll_tuned` component that allow to dynamically set: (1) **the algorithm**; (2) **fan-in/fan-out**; (3) the **segment size**. The main reason for choosing these parameters is that they have been confirmed as having the most impact in several previous studies [38].

The Importance of Tuning MPI Collectives. The importance of this tuning challenge is well known across the MPI community and several studies confirm and further develop the results of our own study: the optimal collective parametrization depends on many factors, such as the physical topology of the system, the number of processes involved, the sizes of the message, as well as the location of the root node [30,39]. An especially thorough analysis of the performance gap between the default parametrization and the optimum parametrization found by exhaustive search for the MPICH implementation is available in [41], and the importance of choosing the right algorithm to perform the collective operations is emphasized in [29]. Using parametrization adapted to the message size and the collective is thus necessary to maximize system's performance. The easiest solution for tuning is to rely on brute force, *i.e.* testing every single possible parametrization and running the benchmark with this parametrization, but this can cause the tuning process to go up to several days in time. Brute force is thus impractical on a production system, as it requires too many computing resources and user time.

All these reasons show the relevance of using black-box optimization through SHAMan: finding the optimal configuration of MPI collective is crucial for the performance of the system as the default parametrization is unsuitable for many communication problems, but exhaustive search is a very impractical way of finding it.

Experiment Plan. For the purpose of demonstrating the efficiency of SHAMan in the case of MPI tuning, we have selected a subset of 4 blocking collectives to tune amongst the most used ones in HPC applications [16,38], and to cover all communication patterns (one-to-all, all-to-one, all-to-all): (1) **Broadcast**, (2) **Gather**, (3) **Reduce** and (4) **Allreduce**.

The tuning is performed using the OSU MPI microbenchmark suite [8], which provides tests for every collective operation. For each of the tuned collective and each tested size, we use the corresponding benchmark in the suite. To ensure stability and reduce the noise when collecting execution times, the OSU benchmark was parameterized to perform 200 warmup runs before performing the actual test. The tuned message size range from 4 KB to 1 MB, with a multiplicative step of 2. Two hardware configurations are selected using the 12 nodes of a cluster, to emulate two of the most common types of process placements encountered in HPC applications: (1) **One MPI process per node**, for a total of 12 MPI processes, as is typical of hybrid applications relying on MPI for inter-node communications and on OpenMP for their implicit, intra-node communications; (2) **One MPI process per core**, for a total of 576 MPI processes, 48 MPI processes per node, which is typical of pure, MPI-only applications which rely on the MPI library for all their communications (inter-node and intra-node alike). This results in a total of 160 SHAMan optimization experiments (4 collectives, 20 sizes and 2 different topologies). The performance metric for tuning is the time elapsed by the benchmark for the selected size of operation, as output by the OSU benchmark.

To provide a thorough analysis of the advantage of black-box optimization compared to exhaustive search, the reference execution time is first computed using the default parametrization which is run 100 times to account for possible noise in the collected execution time. An exhaustive sampling of the parametric space is then performed to select the parametrization with the minimal execution time as the optimal one, which acts as the ground truth. This ground truth is also run 100 times for noise mitigation.

SHAMan Parametrization. The tuning is then performed with SHAMan, using Bayesian Optimization specifically configured for MPI. The initialization plan is composed of 10 parametrization. The selected maximum number of iterations is set to 150 and the stop criterion is improvement based: we choose to stop the optimization process if the best execution time over the last 15 iterations is less than 1% better than the currently found minimum. The best parametrization found by the optimization process is considered to be the best parametrization found by SHAMan and is also run one hundred times to account for noise.

Main Performance Gains

Improvement Compared to Default. The first important result is the gain brought by using the auto-tuner rather than the default parametrization, which is represented in Fig. 7. Over all experiments, we find an average improvement of 48.4% (52.8% in median), using the best parametrization found with Bayesian

Optimization. We find an average improvement of 38.42% (29% in median) for experiments with one mpi process per node and of 58.9% (65.3% in median) when using one mpi process per core, highlighting the efficiency of tuning the Open MPI parametrization instead of simply relying on the default parametrization.

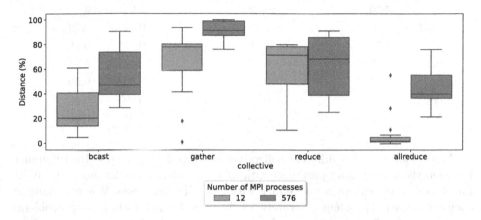

Fig. 7. Performance gain with Bayesian Optimization compared to the default parametrization

The time gain brought by SHAMan varies depending on the tuned collective, as the default parametrization is more adapted than others for some collectives. It is the case for the *allreduce* collective when running one MPI process per node, where the optimum parametrization provides a median improvement of 0.9% (1.8% on average). Other collectives have a default parametrization that is not adapted at all. It is for example the case of the gather collective with one MPI process per core, where we see an improvement of 91% in median and on average. The improvement compared to the default parametrization is strongly dependent on each evaluated parameter: message size, number of processes per node or collectives and is difficult to predict. This highlights the importance of tuning each configuration to get the best performance, and the need for an auto-tuning method that can be used on every architecture.

Tuning Quality Compared to Exhaustive Sampling. The median difference in elapsed time, along with the noise measurement, between the best parametrization found by SHAMan and the optimal parametrization found by exhaustive search is represented in Table 1. Over all optimization experiments, the average distance between the optimum and the result returned by Bayesian Optimization is of 5.71 μs (0.04 in median) for an average noise of respectively 2.03 μs in mean and 0.05 μs in median. This means that in median, the difference between using the best parametrization of our tuner compared to the true best parametrization is imperceptible from the noise. When looking at the relative difference between the optimum and the results from Bayesian Optimization, we find an average distance of 6% (0.7% in median) between the two.

Table 1. Median difference in elapsed time and noise between best parametrization found by Bayesian Optimization and optimal parametrization

Collective	# of MPI processes	Relative difference (%)	ΔT (µs)	Noise (µs)
Allreduce	12	0.51	0.43	0.04
	576	15.28	1.05	3.81
Broadcast	12	4.79	0.32	0.26
	576	2.35	0.32	0.18
Gather	12	0.74	0.04	0.01
	576	0.00	0.02	0.00
Reduce	12	0.00	0.06	0.00
	576	0.00	0.03	0.00

When looking at the different collectives and topologies, we find the difference between the two optimal parametrizations to be inferior to the measured noise for all collectives except for allreduce with 576 MPI processes. When looking at each optimization problem separately, we find that for 105 optimization problems out of 160, the distance of the performance returned by Bayesian Optimization to the optimum is below the measured noise of the system. For the problems where the difference between the results returned by the tuner and the optimum cannot be explained by noise, we find a quite low average difference of 1.90 µs (0.18 in median).

The noise difference between collectives is explained by multiple factors. Gather and reduce show low noise due to their simple communication pattern (all-to-one). On the opposite, the allreduce collective involves much more intertwined messages, which explains its higher noise and noise sensitivity. Broadcast's higher noise is explained by the best performing algorithm found (k-nomial tree) which, according to Subramoni et al. in [37], introduces some noise due the imbalanced communication pattern.

Tuning Speed Compared to Exhaustive Sampling. The elapsed time required to reach the optimum for the two tuning solutions and for each of the collectives and configurations is reported in Table 2.

Table 2. Time to solution for each heuristic and each collective

Collective	# of MPI processes	Exhaustive search (minutes)	SHAMan (minutes)	Gain (%)
Allreduce	12	52.07	4.48	**91.40**
	576	452.55	46.77	**89.66**
Bcast	12	744.10	23.65	96.82
	576	5097.53	133.25	**97.39**
Gather	12	23.45	3.42	**85.42**
	576	1040.07	77.41	**92.56**
Reduce	12	87.50	5.24	**94.01**
	576	550.80	61.58	**88.82**

With a time gain of more than 85% for each collective, we see the benefit of using guided search heuristics with SHAMan to explore the parametric space instead of testing every parametrization with exhaustive sampling. The time required to run all the 160 optimization experiments ranges from a total of 8048 min (approximately 134 h) using brute force to 355 min (approximately 6 h) using Bayesian Optimization, resulting in a total speed-up of 95%. The speed-up is relatively uniform across each collective and each topology.

Overall experiments, we demonstrate that using Bayesian Optimization, we reach 94% of the average potential improvement, for a speed-up in tuning time of 95% on the overall tuning phase. Compared to default Open MPI parametrization, this leads to an average improvement of 48.4% in collective operation performance. This confirms the accuracy of our solution for optimization and makes it a satisfactory alternative to exhaustive search, especially when considering the strong improvement it brings when compared to the default parametrization. This study thus confirms the versatility of SHAMan and black-box optimization for the tuning of a wide range of parametrizable components, and shows that the scope of our work can be extended to many of tunable components within the HPC ecosystem.

6   Conclusion

In this paper, we suggest an OpenSource auto-tuning framework, called SHAMan (**S**mart **HPC** **A**pplication M**a**nager) for tuning noisy and expensive systems, which addresses some of the gaps in the tuning frameworks already present in the literature. While some of our previous works already showed the strong performance of SHAMan on I/O accelerators, we added another use-case to further prove its universality, by tuning MPI collectives. When performing the optimization of four MPI collective communication operations, on two different hardware topologies and for 20 different message sizes, we demonstrate that using Bayesian Optimization, we reach 94% of the average potential improvement, for a speed-up in tuning time of 95% on the overall tuning phase. Compared to default Open MPI parametrization, this leads to an average improvement of 48.4% in collective operation performance. In the near future, we intend to consider additional parameters than the topology and the message size to refine our optimization. Also, we identified additional systems to tune beyond these three cases, such as the MSR parameters of the SAP HANA database which will further extend SHAMan's scope. Regarding the improvement of the optimization methods, we plan on investigating the behavior of the tuner when using pruning strategies. Indeed, these pruning strategies cut off some runs and prevent us from measuring the true performance corresponding to this parametrization. We intend to apply survival analysis to deal with this "censored" data in order to speed up even more SHAMan's convergence speed.

Acknowledgments. This work has been partially funded by the IO-SEA project [4], funded by the European High-Performance Computing Joint Undertaking (JU) and by BMBF/DLR under grant agreement No 955811. The JU receives support from the European Union's Horizon 2020 research and innovation programme and France, the Czech Republic, Germany, Ireland, Sweden and the United Kingdom.

References

1. ARQ. https://arq-docs.helpmanual.io/
2. Atos boosts HPC application efficiency with its new flash accelerator solution. https://atos.net/en/2019/product-news_2019_02_07/atos-boosts-hpc-application-efficiency-new-flash-accelerator-solution
3. Documentation of the SHAMan application. https://shaman-app.readthedocs.io/
4. IO-SEA. https://iosea-project.eu
5. MPICH: a high performance and widely portable implementation of the Message Passing Interface (MPI) standard. https://www.mpich.org/
6. mpitune. https://software.intel.com/content/www/us/en/develop/documentation/mpi-developer-reference-linux/top/command-reference/mpitune.html
7. Open MPI: Open Source High Performance Computing. https://www.open-mpi.org/
8. OSU micro-benchmarks. https://mvapich.cse.ohio-state.edu/benchmarks/
9. Scikit-optimize. https://github.com/scikit-optimize/
10. The SHAMan application. https://github.com/bds-ailab/shaman
11. Tools to improve your efficiency. https://atos.net/wp-content/uploads/2018/07/CT_J1103_180616_RY_F_TOOLSTOIMPR_WEB.pdf
12. Akiba, T., Sano, S., Yanase, T., Ohta, T., Koyama, M.: Optuna: a next-generation hyperparameter optimization framework. In: Proceedings of the 25th ACM SIGKDD International Conference on Knowledge Discovery & Data Mining, pp. 2623–2631 (2019)
13. The GPyOpt authors. GPyOpt: a Bayesian optimization framework in python (2016). https://github.com/SheffieldML/GPyOpt
14. Balaprakash, P., et al.: Autotuning in high-performance computing applications. Proc. IEEE **106**(11), 2068–2083 (2018)
15. Chaarawi, M., Squyres, J.M., Gabriel, E., Feki, S.: A tool for optimizing runtime parameters of open MPI. In: Lastovetsky, A., Kechadi, T., Dongarra, J. (eds.) EuroPVM/MPI 2008. LNCS, vol. 5205, pp. 210–217. Springer, Heidelberg (2008). https://doi.org/10.1007/978-3-540-87475-1_30
16. Chunduri, S., Parker, S., Balaji, P., Harms, K., Kumaran, K.: Characterization of MPI usage on a production supercomputer. In: International Conference for High Performance Computing, Networking, Storage and Analysis, SC 2018, pp. 386–400 (2018)
17. Da Silva, M.D., Tavares, H.L.: Redis Essentials. Packt Publishing (2015)
18. Di Pietro, A., While, L., Barone, L.: Applying evolutionary algorithms to problems with noisy, time-consuming fitness functions. In: Proceedings of the 2004 Congress on Evolutionary Computation, vol. 2, pp. 1254–1261 (2004)
19. Fang, K.T., Li, R., Sudjianto, A.: Design and Modeling for Computer Experiments (Computer Science & Data Analysis). Chapman & Hall/CRC (2005)
20. Faraj, A., Yuan, X.: Automatic generation and tuning of MPI collective communication routines. In: Proceedings of the 19th Annual International Conference on Supercomputing, pp. 393–402 (2005)
21. Hertel, L., Collado, J., Sadowski, P., Ott, J., Baldi, P.: Sherpa: robust hyperparameter optimization for machine learning. In: SoftwareX, vol. 12 (2020)

22. Yoo, A.B., Jette, M.A., Grondona, M.: SLURM: simple Linux utility for resource management. In: Feitelson, D., Rudolph, L., Schwiegelshohn, U. (eds.) JSSPP 2003. LNCS, vol. 2862, pp. 44–60. Springer, Heidelberg (2003). https://doi.org/10.1007/10968987_3

23. Knijnenburg, P., Kisuki, T., O'Boyle, M.: Combined selection of tile sizes and unroll factors using iterative compilation. J. Supercomput. **24**, 43–67 (2003)

24. Koch, P., Golovidov, O., Gardner, S., Wujek, B., Griffin, J., Xu, Y.: Autotune. In: Proceedings of the 24th ACM SIGKDD International Conference on Knowledge Discovery & Data Mining (2018)

25. Le, T.T., Fu, W., Moore, J.H.: Scaling tree-based automated machine learning to biomedical big data with a feature set selector. Bioinformatics **36**, 250–256 (2020)

26. Menon, H., Bhatele, A., Gamblin, T.: Auto-tuning parameter choices in HPC applications using Bayesian optimization. In: 2020 IEEE International Parallel and Distributed Processing Symposium (IPDPS), pp. 831–840 (2020)

27. Merkel, D.: Docker: lightweight Linux containers for consistent development and deployment. Linux J. (239) (2014)

28. Miyazaki, T., Sato, I., Shimizu, N.: Bayesian optimization of HPC systems for energy efficiency. In: Yokota, R., Weiland, M., Keyes, D., Trinitis, C. (eds.) ISC High Performance 2018. LNCS, vol. 10876, pp. 44–62. Springer, Cham (2018). https://doi.org/10.1007/978-3-319-92040-5_3

29. Nishtala, R., Yelick, K.A.: Optimizing collective communication on multicores. In: Proceedings of the First USENIX Conference on Hot Topics in Parallelism (2009)

30. Pjesivac-Grbovic, J., Angskun, T., Bosilca, G., Fagg, G., Gabriel, E., Dongarra, J.: Performance analysis of MPI collective operations. Cluster Comput. **10**, 127–143 (2005)

31. Robert, S., Zertal, S., Goret, G.: Auto-tuning of IO accelerators using black-box optimization. In: Proceedings of the International Conference on High Performance Computing & Simulation (HPCS) (2019)

32. Robert, S.: Auto-tuning of computer systems using block-box optimization: an application to the case of I/O accelerators. Ph.D. thesis, University of UPSaclay (2021)

33. Robert, S., Zertal, S., Couvee, P.: SHAMan: a flexible framework for auto-tuning HPC systems. In: Calzarossa, M.C., Gelenbe, E., Grochla, K., Lent, R., Czachórski, T. (eds.) MASCOTS 2020. LNCS, vol. 12527, pp. 147–158. Springer, Cham (2021). https://doi.org/10.1007/978-3-030-68110-4_10

34. Robert, S., Zertal, S., Vaumourin, G., Couvée, P.: A comparative study of black-box optimization heuristics for online tuning of high performance computing I/O accelerators. Concurrency and Computation: Practice and Experience (2021)

35. Seymour, K., You, H., Dongarra, J.: A comparison of search heuristics for empirical code optimization. In: 2008 IEEE International Conference on Cluster Computing, pp. 421–429 (2008)

36. Siegmund, F., Ng, A., Deb, K.: A comparative study of dynamic resampling strategies for guided evolutionary multi-objective optimization. In: 2013 IEEE Congress on Evolutionary Computation, pp. 1826–1835 (2013)

37. Subramoni, H., et al.: Design and evaluation of network topology-/speed- aware broadcast algorithms for infiniband clusters. In: Proceedings of the IEEE International Conference on Cluster Computing (ICCC), pp. 317–325 (2011)

38. Thakur, R., Rabenseifner, R., Gropp, W.: Optimization of collective communication operations in MPICH. Int. J. High Perform. Comput. Appl. **19**, 49–66 (2005)

39. Tu, B., Zou, M., Zhan, J., Zhao, X., Fan, J.: Multi-core aware optimization for MPI collectives. In: Proceedings of the IEEE International Conference on Cluster Computing, ICCC, pp. 322–325 (2008)
40. Hamadi, Y., Ky, V.K., D'Ambrosio, C., Liberti, L.: Surrogate-based methods for black-box optimization. Int. Trans. Oper. Res. (24) (2016)
41. Vadhiyar, S.S., Fagg, G.E., Dongarra, J.: Automatically tuned collective communications. In: Proceedings of the 2000 ACM/IEEE Conference on Supercomputing, SC 2000 (2000)
42. Zheng, W., et al.: Auto-tuning MPI collective operations on large-scale parallel systems. In: IEEE 21st International Conference on High Performance Computing and Communications, pp. 670–677 (2019)
43. Zielinski, K., Peters, D., Laur, R.: Stopping criteria for single-objective optimization (2005)

Cooperation-Based Search of Global Optima

Damien Vergnet[✉][iD], Elsy Kaddoum, Nicolas Verstaevel,
and Frédéric Amblard

IRIT, Université de Toulouse, CNRS, Toulouse INP, UT3, UT1, UT2,
Toulouse, France
{damien.vergnet,elsy.kaddoum,nicolas.verstaevel,
frederic.amblard}@irit.fr

Abstract. A new cooperation-based metaheuristic is proposed for searching gobal optima of functions. It is based on the assumption that the dynamics of the objective function does not change significantly between iterations. It relies on a local search process coupled with a cooperative semi-local search process. Its performances are compared against four other metaheuristics on unconstrained mono-objective optimization problems. Results show that the proposed metaheuristic is able to find the global minimum of the tested functions faster than the compared methods while reducing the number of iterations and the number of calls of the objective function.

Keywords: Local cooperation · Collective decision · Metaheuristic optimization · Local search

1 Introduction

The simulation of systems is a powerful tool to understand their behaviors and underline their advantages and limits. Several studies aim at reconstructing virtual systems called digital twins to simulate and verify the behavior of specific systems. Such systems can be used in mobility or natural disaster studies to reproduce specific simulation conditions and understand the reasons of such phenomena [4]. Building a digital twin that reproduces the exact behavior of a real system is not an easy task. As real systems are generaly complex systems with non-linear interdependencies among their parameters, finding the best modeling functions and adapting in real-time their parameters to keep a simulation close to the real behavior of the system is not trivial. Many studies have formalised the calibration problem as an optimization problem where the parameters of the modeling functions are tuned by optimizing an objective function: simulation parameters become decision variables and relevent model outputs are integrated into objective functions [2,8]. This implies the need for a fast optimization system that is able to rapidly adapt to changes that may occur in the real system.

B. Dorronsoro et al. (Eds.): OLA 2022, CCIS 1684, pp. 105–116, 2022.
https://doi.org/10.1007/978-3-031-22039-5_9

Multiple optimization methods exist that could be used to solve this problem but they present important drawbacks such as a tendency to converge towards local optima or are too slow [6,11,14].

In this paper we propose a new metaheuristic local optimization method named **CoBOpti**, which stands for **Co**operation-**B**ased **Opti**mization. It is based on an hypothesis of local continuity of the objective function, i.e. the value of the objective function does not vary dramatically when the value of decisions variables varies little. Compared to standard state of the art methods, CoBOpti reaches optimal solutions while reducing the number of iterations and objective function evaluation.

The main contributions of this paper are as follows:

- We introduce a **new local optimization metaheuristic based on an hypothesis of local continuity and cooperation**. This hypothesis allows to model the problem of searching for a global optimum as a **cooperation problem** where a point determines the next point to explore by exploiting the information of its neighbours.
- We experiment and compare our approach on unconstrained mono-objective optimization problems with a single decision variable to demonstrate that **the proposed approach allows to reach a global optimum while minimizing the number of evaluation of the objective function**.

The paper is organised as follow: Sect. 2 discusses the limitations of existing metaheuristics. Section 3 presents our approach and how it gives an answer to these limitations. In Sect. 4, we introduce the results of our experimentation, which is then discussed in Sect. 5 before concluding with limitations and suggest further research.

2   Literature Review

Optimization problems are defined by [3] as finding a vector $\bar{x}_n^* = (x_1^*, ..., x_n^*)$ that optimizes an objective function

$$\bar{f}_k(\bar{x}_n) = (o_1(\bar{x}_n), ..., o_k(\bar{x}_n)) \tag{1}$$

where $\bar{x}_n = (x_1, ..., x_n)$ is a vector of n decision variables.

Many methods exist to solve optimization problems, each making some assumptions on the nature of the problem. One category of such optimization methods is called metaheuristics. [6] defines metaheuristics as methods that perform local and higher level search procedures that are capable of escaping local optima. This definition notably includes methods that employ the notion of neighborhood. The neighborhood of a solution s is the set of all solutions that can be reached from s.

Metaheuristics are interesting for solving optimization problems as they are designed to efficiently explore complex search spaces [6]. Sörensen *et al.* [12] further state that the large majority of real-life optimization problems are more easily solved by metaheuristics, hence our focus on these methods in this paper.

Metaheuristics rely on two important notions: **intensification** and **diversification**. Intensification is a process through which portions of the search space that seem "promising" are explored more thoroughly, i.e. in the neighborhood of the best solutions found yet. Diversification, on the other hand, is a process aimed at exploring unexplored parts of the search space in hopes to find better solutions. It usually relies on a some form of memory of visited solutions [5].

There are numerous metaheuristics, each with their own hypotheses. As the goal of our proposition is to be used to perform on-line calibration, it needs to rely on fast algorithms and to be able to handle the set of visited solutions. The presented methods are thus focused around local search and population-based meta-heuristics.

Local search algorithms explore the search space by exploring the immediate neighborhood of the current solution s and selecting the neighbor solution that has a lower objective value than s. In order to escape from local optima, they feature some sort of hill-climbing process that allows degrading the objective value. Such methods include Simulated Annealing (SA), Generalized Simulated Annealing (GSA), Iterated Local Search, Guided Local Search, etc. [6]. The main advantage of these methods is their rapidity, but an important limitation is their tendency to get stuck in local optima [11]. Somes types of local search metaheuristics rely on some kind of memory of visited solutions to try circumvent this limitation such as Tabu Search [6].

Another category of metaheuristics is the **population-based algorithms**. These methods rely on a set of solutions, called the population. The search space is explored by evaluating each solution and modifying them using a set of simple rules. There are two sub-groups in this category: evolutionary and other nature-inspired methods.

Evolutionary algorithms (EA) are iterative methods centered around the notion of *fitness*. The fitness of a solution represents the quality of this solution based on the objective function. During each iteration, called a *generation*, the fitness of each solution is evaluated. Solutions that feature a high enough fitness value are kept for the next generation, all other are discarded. New solutions are generated by stochasticaly crossing over and modifying (mutating) the solutions that were kept after the selection process. This category includes methods such as Genetic Algorithms, Differential Evolution (DE) and Genetic Programming [6,9]. Contrary to local search methods, EAs explore the search space more thoroughly with bigger population sizes and thus are a lot less susceptible to get stuck in local optima. However, they require more computing power and show slower resolution times.

Other population-based methods behave differently from EAs. They still rely on a set of solutions but draw inspiration from complex biological systems such as bird flocking or ant colonies. They feature the same advantage as EAs, i.e. a more thorough exploration of the search space than local search, but still suffer from the same drawbacks of longer computation times and high computing power requirements [6]. Some methods such as Particle Swarm Optimization (PSO) also suffer from a tendency to converge towards local optima because of a poor distribution of information in the population [14].

In our method we propose to combine the speed of local search approaches and the distribution of information of population-based methods. To achieve this goal we borrow the notions of neighborhood and collective reasoning from these methods. Based on the assumption that **the dynamics of the objective function do not change significantly between two very close points**, we propose a system that **searches for a global optimum through the collective reasoning of already visited solutions**.

Local search and population-based metaheuristics were presented with some of their limitations in the context of optimization for on-line calibration. The next section describes our method, CoBOpti, which is evaluated in Sect. 4.

3 CoBOpti: Cooperation-Based Optimization

In this section, we introduce CoBOpti, a Cooperation-Based Optimization meta-heurtistic. The method we propose combines the advantages of both local search and population-based algorithms: the speed of the former and the information distribution of the latter.

Section 3.1 describes the general principle of the approach by giving an overview of the different search phases; Sect. 3.2 details the local search process; Sect. 3.3 details the semi-local search process and how it enables getting out of local minima; finally, Sect. 3.4 describes how points cooperate to solve specific situations.

3.1 General Principle

The goal of CoBOpti is to iteratively explore the surface of an objective function in order to reach a global optimum. During each iteration, the system has to determine the next point to explore. A **point** p_i is defined as a pair $p_i = (x_i, o_i)$ where x_i is the value of the single decision variable and o_i is the value of the objective function at x_i. The succession of visited points is called a **chain**. The algorithm is composed of 4 phases (Fig. 1).

The algorithm combines two different heursitics: a **local** one (**phases 1, 2 and 3**), which objective is to discover a local minimum, and **semi-local** one (**phase 4**), which uses the set of local minimum already discovered to look for a global minimum.

The goal of **local search (phase 1)** is to find a local minimum. Each iteration t starts with a chain containing some already visited points $p(t)$, $p(t-1)$, etc. Among all the points in the chain, the system choose two points to determine in which direction it needs to go (**phases 2 and 3**). This process continues until a local minimum has been found, i.e. the distance along the x axis between the two points with the lowest objective value of the chain is less than ε_{dist}.

The objective of **semi-local search** is to explore the function towards a global minimum. This process has to decide which point $p(t + 1)$ to explore based on already visited local minima (**phase 4**). Every time the semi-local search has decided on which point to explore next, a new chain is created and the local search continues from this new point.

The search stops when a visited local minimum has an objective value less than a predefined threshold ε_{obj}.

The notion of chains is important as it isolates clusters of points (black and red dots in Fig. 1). It is not desirable that distant points interact during the local search process because of potential higher discrepencies between the actual function value and its estimation. Using chains implies that distant points cannot be used together to compute linear approximations during local search and thus mitigates potential errors. Several chains are created during the optimization process.

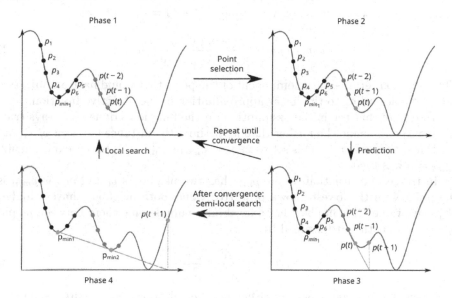

Fig. 1. The search phases of CoBOpti: point selection, local search, higher level search

The following sections detail how points are selected and how $p(t+1)$ is computed. Section 3.2 describes how the local search process selects points to reach a local minima; Sect. 3.3 describes how the system gets out of local minima and searches for a global optimum; finally, Sect. 3.4 describes how points cooperate to solve some difficult situations.

3.2 Local Search

The objective of local search is to follow the curve of the objective function to find a local minimum. At each iteration t, the next point $p(t+1)$ to explore is determined by computing linear approximations of the objective function using two points of the current chain.

Therefore, at each iteration t, two points need to be selected among those in the current chain. The first selected point is the one with the lowest objective value of the chain at time t, noted p_{min}. The second selected point is one of the neighbors of p_{min}. Two points p_1 and p_2 of a chain are said to be **neighbors** if they are immediately next to each other, i.e. there is no third point p_3 between them along the x axis. A point can have a maximum of two neighbors. For example, in Fig. 1, points p_1 and p_2 are neighbors but points p_2 and p_4 are not.

As p_{min} is the point with the lowest objective value of its chain, it has either one or two neighbors at any given time.

Phases 2 and 3 of Fig. 1 illustrate the first situation, where $p(t) = p_{min}$ (green point) has a single neighbor $p(t-1)$. The x component of the next point $p(t+1)$ is computed by a linear approximation of the objective function between $p_{min} = (x_{min}, o_{min})$ and its only neighbor $p(t-1) = p_n = (x_n, o_n)$:

$$x(t+1) = x_n + \frac{-o_n(x_{min} - x_n)}{o_{min} - o_n} \tag{2}$$

This equation returns the x component of the point that would have an objective value of 0 according to the linear approximation of the objective function.

To ensure that the initial assumption on the function's dynamics stays true, the next point cannot be farther than k_{dist} times the distance between p_{min} and p_n. If it is the case, $x(t+1)$ is set to $x_{min} + k_{dist}(x_{min} - x_n)$. In our experiments, $k_{dist} = 5$ was used.

In the second situation, where p_{min} has two neighbors p_l and p_h, as p_{min} is the point with the lowest known objective value, both neighbors have a higher objective value. This implies that a local minimum is somewhere between p_l and p_h. $x(t+1)$ is thus determined by:

$$x(t+1) = \frac{x_{min} + x_n}{2} \tag{3}$$

where x_n is the x component of either p_l or p_h alternatively. Figure 1 shows an example of this situation (black points). The point p_6 was computed this way, using points p_4 as p_{min} and p_5 as its lowest neighbor.

It should be noted that the objective function value does not need to be re-evaluated at the location of the selected neighbor as it is assumed that it has not changed since it was first evaluated.

This whole process repeats until a local minimum is found. The point p_{min} is considered to be a local minimum when the distance to one of its neighbors is less than ε_{dist}.

3.3   Semi-local Search

The goal of the semi-local search is to find a global minimum. The way points are selected is similar to what was described in the local search process but differs in some key aspects.

In order to compute x component of the next point $p(t+1)$ using linear approximations of the objective function, two points are selected: the latest local minimum $p_{min1} = (x_{min1}, o_{min1})$ found by the local search process and one of its neighbors. The **neighbors** of a local minimum are the other adjacent local minima. As with regular points described in Sect. 3.2, local minima have a maximum of two neighbors.

The selected local minimum can have one or two neighbors. Table 1 describes which neighbor is selected depending on the precise situation, where $p_l = (x_l, o_l)$ (resp. $p_h = (x_h, o_h)$) are neighbors of p_{min1} with a lower (resp. higher) x value.

Table 1. Selected neighbor of p_{min1} depending on the situation

	Situation	Selected neighbor
1	One neighbor p_n	p_n
2	Two neighbors, $o_l < o_{min1} < o_h$	p_l
3	Two neighbors, $o_l > o_{min1} > o_h$	p_h
4	Two neighbors, $o_l < o_{min1}$ and $o_{min1} > o_h$	p_l if $o_l < o_h$, otherwise p_h
5	Two neighbors, $o_l > o_{min1}$ and $o_{min1} < o_h$	p_l if $o_l < o_h$, otherwise p_h

For situations 1, 2, 3 and 4, the next point $x(t+1)$ is computed using Eq. 2, swapping p_{min} for p_{min1} and $p(t-1)$ for the selected neighbor. Phase 4 of Fig. 1 illustrates this process for situation 1. In this diagram, there are two known local minima, p_{min1} and p_{min2}, the latter being the newly found one. The next point $p(t+1)$ is estimated using a linear approximation between both local minima. As with the local search, $p(t+1)$ cannot be farther than $k|x_{min1} - x_n|$, if it is the case, the same operations are applied as described in Sect. 3.2.

In situation 5, as both neighbors p_l and p_h of p_{min1} have a higher objective value, a global minimum is probably between p_l and p_h. Equation 3 is used again to determine the next point.

Once $x(t+1)$ has been computed, the local search process resumes from this new point with a new chain.

3.4   Cooperation Mechanisms

Sections 3.2 and 3.3 described the nominal behavior of CoBOpti. The system may encounter a number of special situations during both local and semi-local searches. This section presents cooperation rules to detect and solve them.

Case 1. During local search, when a new chain is created, either because it is the first iteration or the semi-local search created a new one, there is a single point inside the chain. This point thus has no neighbors to compute the next point with. Hence, no linear approximation can be estimated and $x(t+1)$ is directly chosen randomly among $\{x_{min} - \delta, x_{min} + \delta\}$ where $\delta = \frac{1}{k_{prop}}|x_{low} - x_{high}|$ and x_{low} (resp. x_{high}) the lower (resp. higher) bounds of the definition domain of x. In our experiment, $k_{prop} = 100$ was used.

Case 2. During semi-local search, a similar situation may occur where there is only one known local minimum. As there are no neighbors to make linear approximations with, a **hill-climbing** process is initiated to escape the local minimum. This process relies on the two points of the latest chain that have the lowest and highest x value, called extrema. The goal is to climb up the slopes around the local minimum to find another slope of opposite direction.

The search focuses on the slope where the extremum with the lowest objective value is. The next point is computed using Eq. 4 where $p_e^l = (x_e^l, o_e^l)$ is the extremum with the lowest objective value and $p_e^h = (x_e^h, o_e^h)$ is the other. $p_n = (x_n, o_n)$ is the neighbor of p_e^l. This equation computes the x value of the next point which would have an objective value equal to that of the highest extremum, according to the linear approximation of the objective function between the lowest extremum and its neighbor.

$$x(t+1) = x_e^l + \frac{(o_e^h - o_e^l)(x_n - x_e^l)}{o_n - o_e^l} \tag{4}$$

At the next iteration, if the actual objective value is higher than o_e^h, the process switches sides; if this is not the case, it continues as is. This process is repeated until the actual objective value is lower than o_e^l. The local search process then resumes with a new chain.

During this hill-climbing phase, the distance $|x(t+1) - x_e^l|$ cannot be smaller than a threshold δ_{min} in order to prevent the process from slowing down too much.

Case 3. It may happen that the local search process finds a local minimum that was already discovered in previous iterations. In order to escape a potential search loop, two decisions may occur. If a hill-climbing phase was previously initiated at this local minimum, the next point $x(t+1)$ is computed again and multiplied by a factor of 2, to explore twice as far and explore a new area. On the contrary, if no hill-climbing phase was ever initiated at this local minimum, one is started, in hopes to find a new adjacent valley.

Two local minima are considered to be identical if their distance along the x axis is less than a threshold ε_{same}.

In this section we presented our approach. It relies on the notion of chains of points. We first presented a local search process on a chain that allows finding local optima. When a local optimum is found, a semi-local search process allows finding new regions of the search-space to explore. Cooperation mechanisms were introduced to account for special situations, diversify the solutions and create new chains.

In the next section we evaluate the performances of our method. We compare it to four other local-search and population-based metaheuristics on unconstrained mono-objective optimization problems.

4   Experiments and Results

This section compares the performances of CoBOpti with four other methods cited in Sect. 2: Simulated Annealing (SA), Generalized Simulated Annealing (GSA), Differential Evolution (DE) and Particle Swarm Optimization (PSO).

Section 4.1 presents the different test functions used to test the performances; Sect. 4.2 describes the protocole for comparing the performances of CoBOpti with other selected methods; Sect. 4.3 presents the results of the experiments; finally, results are discussed in Sect. 5.

4.1 Test Functions

For the performance comparison experiments, four functions have been selected: Gramacy and Lee (domain: $[0.5, 2.5]$), Ackley (parameters: $d = 1$, $a = 20$, $b = 0.2$, $c = 2\pi$; domain: $[-32, 32]$), Rastrigin (parameter: $d = 1$; domain: $[-5.12, 5.12]$) and Levy function (parameter: $d = 1$; domain: $[-10, 10]$). These functions have been chosen because they feature many local minima, a single global minimum, and a single parameter [1, 7, 10, 13].

4.2 Methods Comparison

The performances of each approach (SA, GSA, DE and PSO) are compared against CoBOpti's. They were all implemented in Python 3.8. GSA and DE were implemented using the `scipy.optimize.dual_annealing` and `scipy.optimize` `.differential_evolution` functions, PSO was implemented with `pyswarm.pso` package, and SA was a custom implementation. For GSA, DE and PSO, all optional parameters excepts those related to bounds, initial state and maximum number of iterations were let to their default value.

Control variables of CoBOpti are set as follows: $\varepsilon_{dist} = 10^{-4}$ (local minimum detection threshold), $\varepsilon_{same} = 0.01$ (minimum distance between local minima), $\delta_{min} = 10^{-4}$ (minimum step size during hill climbing phase), and $\varepsilon_{obj} = 5 \cdot 10^{-3}$ (precision threshold for global minimum objective value).

For every method, except PSO, the initial value v_{init} for each decision variable in a single run is selected by a Sobol Sequence. As values generated by this sequence are all in the $[0, 1]$ interval, they are adjusted to the variable's domain using the formula $v_{init} = s \cdot (d_{max} - d_{min}) + d_{min}$ where s is a value generated by the sequence. We did not specify v_{init} values for PSO as the implementation we used did not allow it.

Three metrics are defined: **success rate**, i.e. the ratio of executions that found the global minimum, **number of iterations, number of evaluations of the objective function**.

4.3 Results

Table 2 shows the success rate, mean number of iterations and function evaluations over 200 executions for each method and function, with a maximum of 1000 iterations.

CoBOpti was able to find the global minimum for all four functions. It took on average between 35 and 100 iterations to find the global minimum with a similar number of objective function evaluations.

The constant 1000 iterations for SA and GSA are explained by their stopping criterion. These methods rely on the number of elapsed iterations to compute probability distributions: the more iterations have passed, the less likely the algorithm is to select a non-improving move. Once the allowed number of iterations has passed, no more non-improving moves can be selected and the algorithm stops. The visited point with the lowest objective value is then returned.

SA did not yield good results, except for Gramacy and Lee's function with nearly 100 % of success rate. It yielded very poor results for Ackley function with only 2 %. These results are coherent with what was described in the review (Sect. 2).

GSA yielded very good results with 100 % on all functions. The number of objective function evaluations was two times higher than SA, around 2000.

DE's success rate is a bit lower than other methods except for SA. However, the mean number of iteration is quite low, staying between 8 and 50.

PSO was able to find the global minimum in all four cases with a low mean number of iterations, between 20 and 50. However, the mean number of function evaluations is higher than other methods, ranging from 2000 to more than 4500.

Table 2. Success rates, average number of iterations and objective function evaluations of tested methods

Method	Function	Success rate	# of iterations	# of evaluations
CoBOpti	G. & L.	100%	49.31	50.31
	Ackley	100%	95.94	96.94
	Rastrigin	100%	80.69	81.69
	Levy	100%	35.3	36.3
SA	G. & L.	99.5%	1000	1000
	Ackley	2%	1000	1000
	Rastrigin	10.5%	1000	1000
	Levy	30%	1000	1000
GSA	G. & L.	100%	1000	2035.58
	Ackley	100%	1000	2124.43
	Rastrigin	100%	1000	2039.97
	Levy	100%	1000	2019.60
DE	G. & L.	97.5%	8.71	154.54
	Ackley	100%	49.62	801.63
	Rastrigin	94%	30.91	481.06
	Levy	100%	50.45	773.75
PSO	G. & L.	100%	20.61	2008.70
	Ackley	100%	46.92	4638.51
	Rastrigin	100%	25.57	2505.57
	Levy	100%	20.02	1951.32

5   Analysis and Discussion

The initial assumption of continuity in function dynamics has been validated by the experiments on several standard functions. CoBOpti showed better success rates than SA and DE, and nearly as good as GSA and PSO. Although the number of iterations of CoBOpti is comparable to that of DE and PSO, its number of function evaluations is several orders of magnitude lower.

This low number of objective function evaluations can be attributed to the fact that the objective function is evaluated only once per visited point. This behavior stems from the initial assumption that states that the dynamics of the objective function does not change significantly between two close points.

Execution times were not shown as differences between methods were not significant. This is most likely due to the relatively low complexity of the selected functions.

A sensitivity analysis should be done to test the influence of k_{dist} and k_{prop} on CoBOpti's performances.

CoBOpti was only tested on mono-objective optimization problems with a single decision variable. Further research is needed to generalize this approach to multi-objective global optimization problems with multiple decision variables. The core principle should stay similar to what was presented in this paper. New cooperation mechanisms should be added to select which objective to minimize and which decision variables to tune at each cycle.

Other experiments could be conducted with other complex functions. As real-world applications are subject to noisy data, resilience to such noise has to be tested.

6   Conclusion

In this paper, CoBOpti, a new metaheuristic for global optimization, was presented. It is based on a hypothesis of local continuity of the dynamics of the objective function. CoBOpti explores the search space by relying on the cooperation of visited solutions based on this hypothesis.

This paper focuses on mono-objective global optimization problems with a single decision variable. Experiments showed that CoBOpti needs less objective function evaluations than other common metaheurstic methods while maintaining similar or better success rates on 1D-functions.

CoBOpti is a promising proposition for use in on-line calibration. Indeed, its low number of objective function evaluations would be useful in the context of on-line calibration of complex simulation models with computationally intensive objective functions. This property could help reduce the time required to calibrate these kinds of models.

References

1. Alauddin, M.: Mosquito flying optimization (MFO). In: 2016 International Conference on Electrical, Electronics, and Optimization Techniques (ICEEOT), pp. 79–84 (2016). https://doi.org/10.1109/ICEEOT.2016.7754783

2. Arsenault, R., Poulin, A., Côté, P., Brissette, F.: Comparison of stochastic optimization algorithms in hydrological model calibration. J. Hydrologic Eng. **19**(7), 1374–1384 (2014). https://doi.org/10.1061/(ASCE)HE.1943-5584.0000938, https://ascelibrary.org/doi/abs/10.1061/(ASCE)HE.1943-5584.0000938

3. Cho, J.H., Wang, Y., Chen, I.R., Chan, K.S., Swami, A.: A survey on modeling and optimizing multi-objective systems. IEEE Commun. Surv. Tutor. **19**(3), 1867–1901 (2017). https://doi.org/10.1109/COMST.2017.2698366

4. Fan, C., Zhang, C., Yahja, A., Mostafavi, A.: Disaster City digital twin: a vision for integrating artificial and human intelligence for disaster management. Int. J. Inf. Manag. **56**, 102049 (2021). https://doi.org/10.1016/j.ijinfomgt.2019.102049, https://www.sciencedirect.com/science/article/pii/S0268401219302956

5. Gendreau, M., Potvin, J.Y.: Tabu search. In: Burke, E.K., Kendall, G. (eds.) Search Methodologies: Introductory Tutorials in Optimization and Decision Support Techniques, pp. 165–186. Springer, Boston (2005). https://doi.org/10.1007/0-387-28356-0_6

6. Gendreau, M., Potvin, J.Y. (eds.): Handbook of Metaheuristics, International Series in Operations Research & Management Science, vol. 146. Springer, Boston (2010). https://doi.org/10.1007/978-1-4419-1665-5, http://link.springer.com/10.1007/978-1-4419-1665-5

7. Gramacy, R.B., Lee, H.K.H.: Cases for the nugget in modeling computer experiments. Stat. Comput. **22**(3), 713–722 (2012). https://doi.org/10.1007/s11222-010-9224-x

8. Ma, J., Dong, H., Zhang, H.M.: Calibration of microsimulation with heuristic optimization methods. Transp. Res. Rec. **1999**(1), 208–217 (2007). https://doi.org/10.3141/1999-22

9. Opara, K.R., Arabas, J.: Differential evolution: a survey of theoretical analyses. Swarm Evol. Comput. **44**, 546–558 (2019). https://doi.org/10.1016/j.swevo.2018.06.010, https://www.sciencedirect.com/science/article/pii/S2210650217304224

10. Potter, M.A., De Jong, K.A.: A cooperative coevolutionary approach to function optimization. In: Davidor, Y., Schwefel, H.-P., Männer, R. (eds.) PPSN 1994. LNCS, vol. 866, pp. 249–257. Springer, Heidelberg (1994). https://doi.org/10.1007/3-540-58484-6_269

11. Storn, R., Price, K.: Differential evolution - a simple and efficient heuristic for global optimization over continuous spaces. J. Glob. Optim. **11**(4), 341–359 (1997). https://doi.org/10.1023/A:1008202821328

12. Sörensen, K., Sevaux, M., Glover, F.: A history of metaheuristics. In: Martí, R., Pardalos, P., Resende, M. (eds.) Handbook of Heuristics, pp. 791–808. Springer, Cham (2017). https://doi.org/10.1007/978-3-319-07124-4_4

13. Valdez, F., Melin, P.: Parallel evolutionary computing using a cluster for mathematical function optimization. In: NAFIPS 2007–2007 Annual Meeting of the North American Fuzzy Information Processing Society, pp. 598–603 (2007). https://doi.org/10.1109/NAFIPS.2007.383908

14. Zhang, Y., Wang, S., Ji, G.: A comprehensive survey on particle swarm optimization algorithm and its applications. Math. Probl. Eng. **2015**, e931256 (2015). https://doi.org/10.1155/2015/931256, https://www.hindawi.com/journals/mpe/2015/931256/

Data-Driven Simulation-Optimization (DSO): An Efficient Approach to Optimize Simulation Models with Databases

Mohammad Dehghanimohammadabadi[✉]

Mechanical and Industrial Engineering Department, Northeastern University,
Boston, MA, USA
m.dehghani@northeastern.edu

Abstract. Simulation-optimization is instrumental to solve stochastic problems with complexity. Over the past half-century, simulation-optimization methods have progressed theoretically and methodologically across different disciplines. The majority of commercial simulation packages - to some degree - offer an optimizer that allows decision-makers to conveniently determine an optimal or near-optimal system design. With the latest advancements in simulation techniques, such as data-driven modeling and Digital Twins, optimizer platforms need a redesign to include new capabilities. This paper proposes a Data-driven Simulation-Optimization (DSO) platform to narrow this gap. By considering data-tables as a decision variable (control), DSO can systematically generate new tables, run experiments, and determine the best table entries to optimize the model. To implement DSO, three software packages (MATLAB, Simio, and MS Excel) are integrated via a customized coded interface, called Simio-API. The applicability of this Simulation-optimization tool is tested in two experimental settings to evaluate its effectiveness and provide some insights for future extensions. The DSO initial results are promising and should stimulate further research in academia and industry.

Keywords: Data-generated modeling · Digital twins · Simheuristics · Intelligent simulation · Simio · Data driven models · Industrial 4.0

1 Introduction

One of the major goals of using simulation modeling is to obtain the ideal configuration of a system. This is achievable by integrating the simulation model with an optimization module. In this approach, the optimizer explores the solution space in order to find the best input values to optimize the model design and its performance measures. A general algebraic form of the simulation optimization (SO) problem can be defined as,

© The Author(s), under exclusive license to Springer Nature Switzerland AG 2022
B. Dorronsoro et al. (Eds.): OLA 2022, CCIS 1684, pp. 117–132, 2022.
https://doi.org/10.1007/978-3-031-22039-5_10

$$\min_{x,y,v} \quad E_v[f(x,y,v)] \tag{1}$$

$$\text{subject to:} \quad E_v[g(x,y,v)] \preccurlyeq 0 \tag{2}$$

$$h(x,y) \preccurlyeq 0 \tag{3}$$

$$x_l < x < x_u \tag{4}$$

$$y_l < y < y_u \tag{5}$$

$$x \in \mathbb{R}^n, y \in \mathbb{D}^m \tag{6}$$

where Eq. (1) refers to a real-valued objective function f evaluation or the model output, which without loss of generality, is minimized. The performance measure of the model is calculated based on the expected output value with respect to input variables. Discrete and continuous input parameters of the simulation model are defined by x and y, respectively. Vector v is the realization of the associated random variables in the model. Equation (2) refers to stochastic constraints where g is a real vector-valued function of stochastic constraints of the model to account for uncertainty. Equation (3) represents a real vector-valued for deterministic constraints h that are not affected by the uncertain parameters. Equations (4) and (5) lower and upper boundaries for discrete and continuous input parameters. Equations (6) defines real and integer decision variables.

The above-mentioned formulation is general enough to represent all kinds of SO problems. Choosing the appropriate simulation and optimization design is crucial for practical applications and highly depends on the problem characteristics [1]. Commercial simulation software packages often use provably convergent algorithms proposed by the research community with metaheuristics to help their users deploy SO for their models [2]. In order to use these optimization algorithms embodied in simulation packages, a user needs to create the preliminary design of the optimization problem. This design includes simulation inputs (controls), the objective functions (responses), and constraints. The optimizer explores a series of simulation configurations by changing the model controls and tries to obtain the optimum or close-to-optimum set of input parameters.

1.1   Research Motivation

Nowadays, using data-table inputs is very essential in developing Discrete-event Simulation (DES) models with large amounts of data. Using data-tables makes simulation model development, execution, and experimentation efficient and easy to implement. Instead of defining an abundant number of parameters and variables, all required data for the simulation modeling can be stored in a data-table format. The data-table values can be manually entered by the user or bound to a data structure framework such as MS Excel, databases, or Enterprise Resource Planning (ERP) systems. This feature provides a highly flexible approach to handling large data inputs for modeling needs. Data-tables could efficiently include any type of model's information with a large number of entries. This could include entities' information (e.g. entity types, arrival times, processing time(s), sequence and priorities, etc.), resources' data (e.g. schedules, maintenance plans, locations on the layout, etc.), or even transportation networks. Based on the new advancement in DES commercial packages, these data can be accessed sequentially, randomly, directly, and even automatically [3]. This emphasizes the importance of using data-tables with the simulation model creation and experimental analysis.

1.2   Novelty and Main Contributions of the Paper

Although enhancing a simulation model with data-table inputs simplifies the model development, there is not a trivial way to optimize the simulation models based on data-table inputs. Existing commercial SO tools are designed to include a limited number of numerical controls (binary, integer, or real) for optimization purposes and are incapable of including data-table inputs to the optimization process. This becomes more challenging when data-tables are non-numerical, e.g. categorical, Date/Time, etc. Therefore, this paper aims to remedy this lack and introduce Data-Table Simulation-Optimization (DSO) platform.

This platform that allows simulation users to optimize the model's performance by deploying simulation experiments with respect to data-table inputs with no data format restrictions. For instance, a user can optimize "Patient's Arrival Table" (Data/Time format) in a healthcare system, "Product Mix Table" (categorical format) in a manufacturing setting, or "Destinations/Nodes Sequence Table" (integer format) for a set of transporters/vehicles in a supply chain network. This platform benefits simulation model extensibility, scenario creation, and experiment repeatability. So, the original form of the SO problem can be modified as follows:

$$\min_{x,y,t,v} \quad E_v[f(x,y,t,v)] \tag{7}$$

$$\text{subject to:} \quad E_v[g(x,y,t,v)] \preceq 0 \tag{8}$$

$$h(x,y,t) \preceq 0 \tag{9}$$

$$x_l < x < x_u \tag{10}$$

$$y_l < y < y_u \tag{11}$$

$$x \in \mathbb{R}^n, y \in \mathbb{D}^m \tag{12}$$

$$t \in \mathbb{T} \tag{13}$$

where Eqs. 7–11 are equivalent of Eqs. 1–5. The only difference is adding a new term t to represent data-table inputs used towards simulation optimization. The values stored in the table could have any format and structure, where all of these formats are represented by T (Eq. 13). The proposed DSO platform introduces the following capabilities:

- **Data-driven:** the ability to include data-table inputs as a decision variable (control) in the optimization process. The optimizer automatically generates different values across the entire table, and gradually evolves its values to optimality.

- **Generalized:** the ability to handle multiple data types for optimization purposes. This includes but not limited to dates (e.g. order dates, dues dates, lead times), strings (e.g. dispatching rules name), numeric (e.g. resource capacities, order quantities, entity priority), objects list (e.g. list of nodes, list of tasks, list of transporters).
- **Scalable:** optimize a large set of parameters simultaneously. For instance, to optimize priority values of 100 jobs with 2 servers, the existing solution approaches require an exhaustive list of parameters (in this cases $(100)(2) = 200$) which is which is neither practical nor scalable. By considering the table as one-single input, DSO reduces the need of defining a large number of individual parameters. This makes the optimization process scalable, hassle free, and easy to implement.
- **Real-time:** leveraging data-tables enables the simulation model to be connected with a stream of data extracted from ERP systems. This allows DSO to be more aware of the changes in the real system and provide dynamic and reliable results.

1.3 Organization and Structure of the Paper

The rest of this paper is organized as follows. Section 2 provides a brief introduction of SO and highlights the motivation of this work. In Sect. 3, the DSO platform is explained in detail and its implementation aspects are discussed. To demonstrate the applicability of the proposed framework, two experiments are designed and analyzed in Sect. 4. This work is wrapped up in Sect. 5, and future work directions are presented in the end.

2   Literature Review and Background

The desirability of seeking better solutions is the main driver for developing SO techniques. As a definition, SO is a systematic search process to find the best configuration of a stochastic system in order to optimize objective function(s). This search needs to be efficient to minimize the resources spent while maximizing the information obtained in a simulation experiment [4]. With a long and illustrious history, SO is arguably the ultimate aim of most simulationists [5]. With a huge advancement in both research and practice, SO is considered as one of the main streams of simulation studies and has received considerable attention from both simulation researchers and practitioners [6]. Many studies applied SO to address problems in healthcare [7–9], manufacturing [10–12], and supply chain [13–16].

Nowadays, SO is a vibrant field and various sub-disciplines are evolved from different communities such as systems and control, statistics and design-of-experiments, math programming, and even computer science [5]. With the advancements in the SO literature, many of the simulation vendors provide some sort of automatic experimental generators or optimization tools. Based on a survey conducted by [17], 40 out of 55 software packages are featured with SO tools. Most of these SO tools such as OptQuest [18] and SimRunner [19] are designed based on the Simheuristics structure. In Simheuristics, a Metaheuristic algorithm is coupled with the simulation environment [20] to perform an iterative search through parameter values to obtain improved responses. These tools do not offer guarantees of optimality, but the provided solutions are near-to-optimal and realistic.

Almost in all the existing SO tools, the decision variables are restricted to numerical parameters without considering other important elements of the model. For instance, in a healthcare system, one can easily evaluate a hospital model based on different numbers of resources (i.e. physicians, nurses, beds, etc.) and determine the best numeric combination of these values. However, within the same model, it is not trivial to systematically change the layout of the hospital, the schedule of physicians/nurses, or even individual patients' arrival times. To evaluate each of the above-mentioned scenarios, a tremendous amount of effort is required to manually make changes and customize the simulation model. The same concept relates to a manufacturing setting. For example, finding the best number of workers, transporters, servers, etc. is easily attainable using the existing commercial or non-commercial SO tools, while optimizing workers' schedules, transporters' network, and servers' location (layouts) is not a straightforward task.

The proposed DSO in this article is a perfect alternative in which numerical and data-table inputs can be optimized simultaneously. This platform is general enough to include any type of table entries with multiple data formats. Therefore, DSO is a promising SO framework and can introduce a significant opportunity for the simulation community to solve problems efficiently on a larger scale.

3   The Proposed DSO Platform

To obtain a practical and ideal DSO model, an integrated framework is developed using three modules, namely (i) optimization, (ii) simulation, and (iii) data exchange. As illustrated in Fig. 1, both simulation and optimization modules are bound to an external data source. In this iterative scheme, the optimization module provides a new solution and updates the data-table input(s) in each iteration, and then, the simulation module runs the model based on the newly provided data settings. Using this new framework, users can easily build a simulation model with data-tables and optimize it in an efficient manner.

The main goal is to make this framework general enough to be used in any simulation setting with different applications. To implement the proposed DSO framework, three software packages are linked together. The Optimization Module is deployed in MATLAB, and the Simulation Module is designed in Simio. These two modules are linked together via MS Excel for data-table input exchange.

MATLAB is a powerful tool and is adequate for addressing heavy computing needs such as optimization problems. MATLAB can be easily linked with

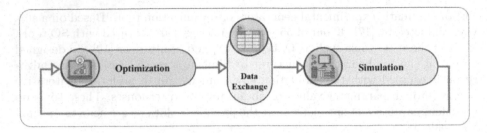

Fig. 1. The proposed DSO platform structure.

external software for advanced calculations and its programmable environment allows users to design and customize their algorithms. It has enhanced optimization capabilities and allows its user to choose between the existing optimization libraries or their developed algorithms. As a result, the applied optimization algorithm in this work is coded in MATLAB.

Simulation models often require large amounts of data to define different elements of the model such as entities, objects, networks, schedules, etc. Simio can represent data in simple tables and allows users to match the data schema for the manufacturing data (e.g. an ERP system) [21]. This simulation software is flexible enough to model complex systems with different operational needs. Another major benefit of Simio is its API capability which helps developers to extend Simio's access to external software packages. By taking advantage of this feature, a customized API is coded in C# to assist with the TDSO idea. This API connects MATLAB with Simio and provides a scheme to exchange data between the two. This enables users to connect Simio with other programming languages such as Python, R, or Julia.

The third component of this framework is MS Excel which is compatible with both MATLAB and Simio. This Data Exchange Module is the central piece of the framework and facilitates the data transfer between simulation and optimization packages. In Simio, data-tables can be bound to MS Excel and be accessed sequentially, randomly, directly, and even automatically. This important feature makes the development of DSO a feasible and effective intervention. A schematic illustration of DSO is shown in Figure 2 and its pseudocode is provided in Algorithm 1.

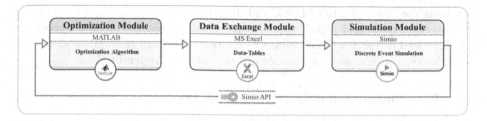

Fig. 2. Implementation components of the DTSO framework using MATLAB, Simio, and Excel.

Algorithm 1: Data-driven Simulation Optimization (DSO)

Data: Define the simulation model parameters: $x \in \mathbb{R}^n$, $y \in \mathbb{D}^m$
Data: Define the simulation model data-table inputs $t \in T$

initialization;
while *not the end of optimization algorithm* **do**

    Generate a new data-table solution
    Update data with the new data-table solution
    Trigger the simulation model and run experiments
    for $r \leftarrow 1$ **to** *MaxReplications* **do**
        Replicate the simulation model
        Calculate the expected value of simulation responses
        $\mathbb{E}_v[f(x, y, t, v)]$
        Update objective function values

Result: Optimal/near-to-optimal data-table input

---

4 Experimental Analysis: DSO for Job Scheduling and Sequencing

Using data-tables can significantly facilitate simulation modeling development, execution, and improvement. Data can be imported, exported, and even bound to external resources. While reading and writing disk files interactively during a run can reduce the execution speed, tables hold their data in memory and so execute very quickly [22]. Applying DSO can harness the advantages of data-tables and place more emphasis on their use. To reveal some insights into the present and future works, this section demonstrates the applicability of DSO in two experimental settings.

To maintain the focus of the paper on the DSO advantages, the following experiments in this section introduce typical manufacturing settings with nominal operations. However, without loss of generality, DSO can be utilized in any simulation models in Simio with different levels of complexities. Also, the applied optimization algorithm is Particle Swarm Optimization (PSO) which is manually coded in MATLAB to optimize data-table entries. Again, this does not limit the applicability of DSO, and different users can leverage a variety of optimization tools and algorithms to solve their problems. These experiments are discussed as follows. To maintain the paper's flow and readability, details of PSO, its operations, and pseudocode are provided in Appendix A.

4.1 Experiment 1: Job Scheduling with DSO

This study considers a flow shop model where 50 jobs (entities) are processed sequentially by two (2) servers. This model includes three (3) types of jobs that randomly arrive in the system in batches of five (5). The model assumptions are:

- All machines are ready to be scheduled in time zero.
- Preemption of operations of each job is not allowed.
- Different job types have different distributions for processing time and due dates.
- Setup time is job dependent (setup time varies from one job type to another on each server.)
- Each machine can process only one operation at a time.

Figure 3-a depicts the simulation environment where the flow shop model is developed based on the assumptions provided above. This model has two objective functions: i) minimizing the average Time In the System (TIS), and ii) minimizing the Total Tardiness Cost (TTC) of all jobs. The goal is to find the best prioritization of jobs in both servers in such a way that all objective functions are optimized. This model is simulated in Simio and a data-table is created to implement its operational logic.

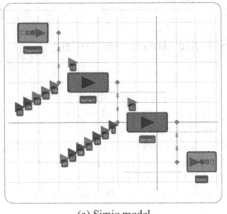

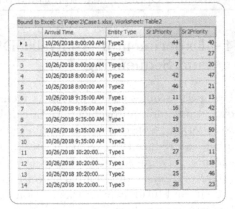

	Arrival Time	Entity Type	Sr 1Priority	Sr 2Priority
▶ 1	10/26/2018 8:00:00 AM	Type2	44	40
2	10/26/2018 8:00:00 AM	Type3	4	27
3	10/26/2018 8:00:00 AM	Type1	7	20
4	10/26/2018 8:00:00 AM	Type2	42	47
5	10/26/2018 8:00:00 AM	Type2	46	21
6	10/26/2018 9:35:00 AM	Type1	11	13
7	10/26/2018 9:35:00 AM	Type3	16	42
8	10/26/2018 9:35:00 AM	Type1	19	33
9	10/26/2018 9:35:00 AM	Type3	33	50
10	10/26/2018 9:35:00 AM	Type2	49	48
11	10/26/2018 10:20:00...	Type1	27	11
12	10/26/2018 10:20:00...	Type1	5	18
13	10/26/2018 10:20:00...	Type2	25	46
14	10/26/2018 10:20:00...	Type3	28	23

Bound to Excel: C:\Paper2\Case1.xlsx, Worksheet: Table2

(a) Simio model   (b) Jobs' priority in Server1 and Server2

Fig. 3. Job Scheduling simulation model and the data-table input structure.

Figure 3-b depicts a snapshot of the data-table entry which stores entities' information such as arrival time, entity type, and priority numbers in Servers 1 and 2. The highlighted columns (Sr1Priority and Sr2Priority) indicate the priority of jobs on each server. This Optimization Module (i.e. PSO algorithm) in DSO treats this table as a decision variable (control) and improves its entries sequentially until the desired solution is obtained. In every iteration, the optimization module in DSO changes job priorities (values in Sr1Priority and Sr2Priority columns). Then, the simulation is triggered to simulate the model based on new data-table entries and runs replications to calculate the expected value of objective functions. These results are transferred to the Optimization Module to generate new solutions. This cycle repeats until the stopping criteria (which are usually set by the user) are met.

To analyze the performance of DSO, its results are compared with two heuristic methods available in the literature for solving flow shop scheduling.

- **Heuristic 1- SPT:** Shortest Processing Time or SPT has shown superior performance for job scheduling in many research investigations [23]. By ranking jobs based on the ascending order of their processing times, SPT minimizes the total completion time.
- **Heuristic 2- EDD:** The second heuristic is EDD (Earliest Due Date) which arranges job orders to minimize the total tardiness cost of jobs [24].

The applied PSO algorithm in DSO uses a weighted average of both objective functions to solve the problem (Eq. 14).

$$\min_{x,y,t,v} \quad w_1 \times \mathbb{E}_v\left[TIS\left(x,y,t,\ v\right)\right] + w_2 \times \mathbb{E}_v\left[TTC\left(x,y,t,\ v\right)\right] \qquad (14)$$

The performance of the calculated optimal solution provided by DSO is compared with heuristic results provided by SPT and EDD. Each of these solutions is replicated 200 times and the ultimate results are plotted in Fig. 4. These results suggest the superiority of DSO over SPT and EDD in terms of both objective functions (time in the system and tardiness cost). With a smooth and straightforward implementation, DSO could efficiently improve the prioritization of jobs and provide competitive results.

Insight 1: The applicability of DSO can be easily extended to a flow shop model with more servers. To include a new server in the simulation model, a new column needs to be added to the data-table to represent jobs' priority on that server (i.e. Sr3Priorirty). In the optimization algorithm, the size of decision variables ($nVar$) directly depends on the number of jobs (n) and the number of servers m in the model ($nVar = n \times m$). So, adding a new server or more jobs just needs a slight change in the optimization algorithm and updating $nVar$ parameter.

4.2   Experiment 2: Job Sequencing with DSO

The second experiments demonstrate the applicability of DSO in solving a job sequencing problem in a multi-stage flow shop system (known as assembly flow

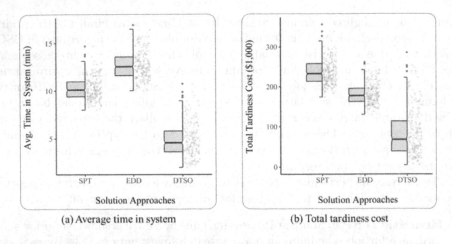

(a) Average time in system   (b) Total tardiness cost

Fig. 4. Experimental results of the flow shop scheduling study.

shop). In this case, 90 jobs are released to the floor with 3 stages and 5 servers in each. The model assumptions are:

- Each machine can process only one operation at a time.
- Assembly or post-processing stages begin readily after all previous stage operations are completed.
- All machines are ready to be scheduled in time zero.
- Preemption of operations of each job is not allowed.
- The setup time is zero.

Figure 5 depicts the simulation environment where this assembly shop is developed. Objective functions are to minimize both i) average Time In the System (TIS), and ii) Total Tardiness Cost (TTC) of all jobs. In all stages, the processing time of jobs is set to $normal(20, 2)$ min, and due dates are uniformly distributed using $uniform(50, 250)$ min. There are five (5) homogenous servers

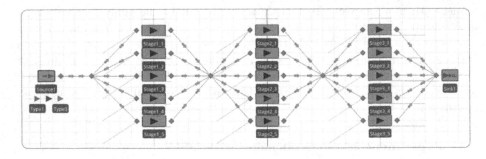

Fig. 5. Multi-stage flow shop system developed in Simio.

in each stage and each job has to follow a sequence of tasks to complete the assembly.

The goal is to find the best sequence for each job in each stage. In other words, the ideal solution should determine which server is selected in each stage to process a given job. To deploy this, a data-table is utilized in the simulation model to set up the sequencing logic. Figure 6 shows a screenshot of this table where each row represents a given job arrival time and its sequence in different stages. Three highlighted columns (Stage1Sr, Stage2Sr, and Stage3Sr) indicate the server IDs (1 to 5) that each job has to go through sequentially from stage 1 to 3 before leaving the system.

Bound to Excel: Case2.xlsx, Worksheet: Table2
Data has not been imported or is of an unknown age

	Arrival Time	Stage 1Sr	Stage 2Sr	Stage 3Sr
▶ 1	10/26/2018 8:00:00 AM	3	3	5
2	10/26/2018 8:00:00 AM	1	3	2
3	10/26/2018 8:00:00 AM	2	3	3
4	10/26/2018 8:00:46 AM	4	5	5
⋮	⋮	⋮	⋮	⋮
89	10/26/2018 1:31:27 PM	3	5	3
90	10/26/2018 1:32:37 PM	1	4	4

Fig. 6. Data-table input structure for the job sequencing problem.

To optimize this problem, the DSO platform explores different combinations of sequences for all jobs and provides a solution that minimizes objective functions. Two heuristics rules are considered to evaluate the quality of DSO results. These heuristics are:

- **Heuristic 1- Cyclic:** In each stage, this rule selects servers cyclically to carry out new jobs.
- **Heuristic 2- LLS:** This routing rule, selects a server with the lowest service load (LLS) upon a new job arrival to the stage.

The obtained solutions of these three approaches are simulated with 200 replications to estimate the expected value of objective functions, TIS, and TTC. The jitter-boxplots of these results are presented in Fig. 7, where DSO performance is adequately better than heuristics. The optimization algorithm in DSO could find a solution with higher quality and less variability. The significance of this difference is tested using one-way ANOVA for TIS and TTC objectives (Table 1 and Table 2). The p-value of both tests is low ($p < 0.001$), which appears that DSO's superiority is statistically significant.

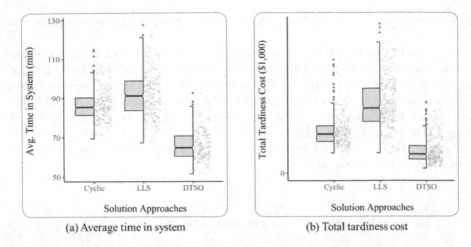

(a) Average time in system   (b) Total tardiness cost

Fig. 7. Data-table input structure for the job sequencing problem.

Table 1. One-way ANOVA results for the average time in the system results.

Source of variation	SS	df	MS	F	P-value	F critical
Between groups	76580.58	2	38290.29	474.10	5.2E−124	3.010815
Within groups	48215.91	597	80.76			
Total	124796.5	599				

Table 2. One-way ANOVA results for the total tardiness cost.

Source of variation	SS	df	MS	F	P-value	F critical
Between groups	2.63E+08	2	1.32E+08	309.3843	6.31E−93	3.010815
Within groups	2.54E+08	597	425708.2			
Total	5.18E+08	599				

Insight 2: In this experiment, DSO solved the problem of sequencing for 90 jobs in 3 stages ($nVar = 270$). To solve this problem, OptQuest requires at least 180 controls (properties) to find the solution; whereas, DSO takes the data-table as a decision variable and evolves its values until the desired solution is achieved.

5   Conclusion and Future Works

For more than fifty years, simulation has been extensively applied by researchers and developers. Traditionally, it is appealing to equip simulation with optimization to tackle stochasticity and the complexity of problems. Despite tremendous advancements in SO techniques, designing well-established models is still demanding by today's standards [1]. In addition, simulation software packages fail to incorporate human decision-making analysis or highly computational support tools [25].

This article introduced an innovative interaction between simulation and optimization to carry out the decision-making process based on table-table inputs. Unlike the existing commercial software packages, the proposed DSO framework can efficiently solve problems with a large set of parameters and data-tables. By developing an application programming interface, called Simio-API, this framework connects three modules, simulation, optimization, and data exchange. This integration can run simulation experiments with multiple data-table settings, evolve their values, and achieve satisfactory results.

The usefulness of the proposed framework is demonstrated in two experimental scenarios, 1) job scheduling in a flow shop system, and 2) Job sequencing in a multi-stage flow shop system. These experiments demonstrated DSO's applicability and efficiency. More importantly, some insights are provided to show how the new model can be implemented and extended to new works.

Introducing DSO offers new opportunities to the community and paves a new avenue of research for theory and practice. Nowadays, many simulation software developers value data-driven models. With the emergence of new simulation techniques such as data-generated modeling and Digital Twins, the usefulness of DSO can become more obvious. The concept of data-generated modeling is based on creating a simulation model automatically using data-tables. This populates a complete simulation model from scratch by adding objects to the environment and mapping them to the tables. These tables can include all of the modeling needs such as resources' information, entities, networks, transports, schedules, tasks, etc. Lately, Simio announced a newly added feature to its software to build data-generated models [26]. By leveraging DSO, one can optimize different components of the model systematically without the need for manual changes. For instance, optimizing system layouts (i.e. hospitals, manufacturing systems, etc.) are traditionally limited to a few scenarios suggested by layout designers. By taking advantage of data-generated modeling in Simio and DSO, one can easily change the layout, make simulation models instantly, and experiment with an abundance of layouts. As shown in Fig. 8-a, this requires defining object coordinates as decision variables (XLocation and ZLocation columns) for DSO and let it solve the problem. Another example could be optimizing manufacturing orders where orders' release date and priority need to be optimized (Fig. 8-b).

Another future research direction is to explore other optimization techniques in the model. To keep the focus of the paper on the platform, just one algorithm (PSO) is used in experiments. However, there are plenty of SO techniques that can be borrowed and tested within DSO design. Multiple metaheuristic algorithms can be utilized to develop data-table Simheuristics and solve the problems. By increasing the size of data-tables (and decision variables), neural networks can be added to the optimizer and make the model computationally efficient by approximating objective functions.

(a) Resources data-table   (b) Manufacturing orders data-table

Fig. 8. Examples of data-generated modeling in Simio.

A   Appendix A

Particle Swarm Optimization or PSO is a population-based Metaheuristic algorithm developed by Kennedy and Eberhart in 1995 [27] PSO is a swarm-based algorithm and by moving particles in a specific exploration field [28]. Due to its effective balancing of exploration and exploitation [29], PSO has been widely used in the development of Simheuristic models and in solving SO problems. Recent examples include using PSO to deal with stochastic models in supply chain management [30], healthcare systems [31], and manufacturing [32]. The general pseudocode of PSO is shown in Algorithm 2.

Algorithm 2: Pseudo-code of Particle Swarm Optimization (PSO)

1: // **Initialization**   ▷ Generate particles

  for $i = 1 : N_s$ **do**

2:

  Initialize $s_i(t = 0)$

3: Initialize $v_i(t = 0)$

4: $P_i^{best} \leftarrow s_i$

5: // **PSO loop**

6: $G^{best} = 0$ **for** $t = 1 : Max_{it}$ **do**

  for $i = 1 : N_s$ **do**

7:

  $$v_i(t+1) = wv_i(t) + c_1r_1[P_i^{best} - s_i(t)] + c_2r_2[G^{best} - s_i(t)]$$

8: $s_i(t+1) \leftarrow s_i(t) + v_i(t+1)$

9: Evaluate $fitness\, s_i(t+1)$

  if $fitness(P_i^{best}) < fitness(s_i(t+1))$ **then**

10:

  $P_i^{best} \leftarrow s_i(t+1)$ ▷ Update Personal best **if** $fitness(G^{best}) < fitness(P_i^{best})$

  then

11:

  $G^{best} \leftarrow P_i^{best}$ ▷ Update Global best

12: $t \leftarrow t + 1$

Notations:

N_s : Swarm size, $i = 1, 2, ..., N_s$: Particles index

s_i : Solution (particle), v_i : Velocity, w : Inertia weight

P_i^{best} : Personal best solution, c_1 : Personal learning factor

G^{best} Global best, c_2 : Global learning factor

r_1, r_2 : Random numbers $\sim u(0, 1)$

MAX_{it} : Max number of iterations, $t : 0, 1, 2, ..., MAX_{it}$: Iteration index

---

References

1. Figueira, G., Almada-Lobo, B.: Hybrid simulation-optimization methods: a taxonomy and discussion. Simul. Modell. Pract. Theory Simul.-Optim. Complex Syst.: Methods Appl. **46**, 118–134 (2014). ISSN 1569-190X. https://doi.org/10.1016/j.simpat.2014.03.007, http://www.sciencedirect.com/science/article/pii/S1569190X14000458. Accessed 29 May 2016
2. Amaran, S., Sahinidis, N.V., Sharda, B., Bury, S.J.: Simulation optimization: a review of algorithms and applications. 4OR **12**(4), 301–333 (2014). http://link.springer.com/article/10.1007/s10288-014-0275-2. Accessed 01 May 2017
3. Smith, J.S., Sturrock, D.T., Kelton, W.D.: Simio and Simulation: Modeling, Analysis, Applications: 4th Edition - Economy, English, 4 edn. CreateSpace Independent Publishing Platform (2017). ISBN 978-1-5464-6192-0
4. Carson, Y., Maria, A.: Simulation optimization: methods and applications. In: Conference Proceedings, pp. 118–126. IEEE Computer Society (1997)
5. Fu, M.C., Henderson, S.G.: History of seeking better solutions, AKA simulation optimization. In: 2017 Winter Simulation Conference (WSC), pp. 131–157. IEEE (2017)
6. Ólafsson, S., Kim, J.: Simulation optimization. In: Proceedings of the Winter Simulation Conference, vol. 1, pp. 79–84. IEEE (2002)
7. Dehghanimohammadabadi, M., Kabadayi, N.: A two-stage AHP multi- objective simulation optimization approach in healthcare. Int. J. Anal. Hierarchy Process **12**(1), 117–135 (2020)
8. Azadeh, A., Ahvazi, M.P., Haghighii, S.M., Keramati, A.: Simulation optimization of an emergency department by modeling human errors. Simul. Modell. Pract. Theory **67**, 117–136 (2016)
9. Rezaeiahari, M., Khasawneh, M.T.: Simulation optimization approach for patient scheduling at destination medical centers. Expert Syst. Appl. **140**, 112 881 (2020)
10. Seif, J., Dehghanimohammadabadi, M., Yu, A.J.: Integrated preventive maintenance and flow shop scheduling under uncertainty. Flex. Serv. Manuf. J. **32**, 852–887 (2020). https://doi.org/10.1007/s10696-019-09357-4
11. Aiassi, R., Sajadi, S.M., Molana, S.M.H., Babgohari, A.Z.: Designing a stochastic multi-objective simulation-based optimization model for sales and operations planning in built-to-order environment with uncertain distant outsourcing. Simul. Modell. Pract. Theory **104**, 102103 (2020)
12. Amiri, F., Shirazi, B., Tajdin, A.: Multi-objective simulation optimization for uncertain resource assignment and job sequence in automated flexible job shop. Appl. Soft Comput. **75**, 190–202 (2019)
13. Drenovac, D., Vidović, M., Bjelić, N.: Optimization and simulation approach to optimal scheduling of deteriorating goods collection vehicles respecting stochastic service and transport times. Simul. Modell. Pract. Theory **103**, 102 097 (2020)
14. Kabadayi, N., Dehghanimohammadabadi, M.: Multi-objective supplier selection process: a simulation-optimization framework integrated with MCDM. Ann. Oper. Res. **319**, 1607–1629 (2022). https://doi.org/10.1007/s10479-021-04424-2
15. Vieira, A.A., Dias, L., Santos, M.Y., Pereira, G.A., Oliveira, J.: Are simulation tools ready for big data? Computational experiments with supply chain models developed in Simio. Proc. Manuf. **42**, 125–131 (2020)
16. Goodarzian, F., Hosseini-Nasab, H., Muñuzuri, J., Fakhrzad, M.-B.: A multi-objective pharmaceutical supply chain network based on a robust fuzzy model: a comparison of meta-heuristics. Appl. Soft Comput. **92**, 106 331 (2020)

17. Swain, J.J.: Simulated worlds. OR/MS Today **42**(5), 36–49 (2015)
18. Laguna, M.: Optimization of Complex Systems with OptQuest. A White Paper from OptTek Systems Inc. (1997)
19. Hein, D.L., Harrell, C.R.: Simulation modeling and optimization using ProModel. In: 1998 Winter Simulation Conference. Proceedings (Cat. No. 98CH36274), vol. 1, pp. 191–197. IEEE (1998)
20. Juan, A.A., Faulin, J., Grasman, S.E., Rabe, M., Figueira, G.: A review of simheuristics: extending metaheuristics to deal with stochastic combinatorial optimization problems. Oper. Res. Perspect. **2**, 62–72 (2015)
21. Xu, J., Huang, E., Hsieh, L., Lee, L.H., Jia, Q.-S., Chen, C.-H.: Simulation optimization in the era of industrial 4.0 and the industrial internet. J. Simul. **10**(4), 310–320 (2016)
22. Jian, N., Freund, D., Wiberg, H.M., Henderson, S.G.: Simulation optimization for a large-scale bike-sharing system. In: 2016 Winter Simulation Conference (WSC), pp. 602–613. IEEE (2016)
23. Pegden, C.D.: Introduction to SIMIO. In: 2008 Winter Simulation Conference, pp. 229–235. IEEE (2008)
24. Sturrock, D.T.: Traditional simulation applications in industry 4.0. In: Gunal, M.M. (ed.) Simulation for Industry 4.0. SSAM, pp. 39–54. Springer, Cham (2019). https://doi.org/10.1007/978-3-030-04137-3_3
25. Jules, G., Saadat, M., Saeidlou, S.: Holonic goal-driven scheduling model for manufacturing networks. In: 2013 IEEE International Conference on Systems, Man, and Cybernetics, pp. 1235–1240. IEEE (2013)
26. Dehghanimohammadabadi, M.: Iterative optimization-based simulation (IOS) with Predictable and unpredictable trigger events in simulated time. Ph.D. thesis, Western New England University (2016). http://gradworks.umi.com/10/03/10032181.html. Accessed 30 May 2016
27. Dehghanimohammadabadi, M., Keyser, T.K.: Intelligent simulation: integration of SIMIO and MATLAB to deploy decision support systems to simulation environment. Simul. Modell. Pract. Theory **71**, 45–60 (2017). http://www.sciencedirect.com/science/article/pii/S1569190X16301356. Accessed 17 Dec 2016
28. Sturrock, D.T.: Using commercial software to create a digital twin. In: Gunal, M.M. (ed.) Simulation for Industry 4.0. SSAM, pp. 191–210. Springer, Cham (2019). https://doi.org/10.1007/978-3-030-04137-3_12
29. Kennedy, J., Eberhart, R.: Particle swarm optimization. In: Proceedings of ICNN 1995-International Conference on Neural Networks, vol. 4, pp. 1942–1948. IEEE (1995)
30. Shaheen, M.A., Hasanien, H.M., Alkuhayli, A.: A novel hybrid GWOPSO optimization technique for optimal reactive power dispatch problem solution. Ain Shams Eng. J. **12**, 621–630 (2020)
31. Usman, M., Pang, W., Coghill, G.M.: Inferring structure and parameters of dynamic system models simultaneously using swarm intelligence approaches. Memetic Comput. **12**(3), 267–282 (2020)
32. Park, K.: A heuristic simulation-optimization approach to information sharing in supply chains. Symmetry **12**(8), 1319 (2020)

Logistics

Sweep Algorithms for the Vehicle Routing Problem with Time Windows

Philipp Armbrust[1]([⊠]), Kerstin Maier[1], and Christian Truden[2]

[1] Department of Mathematics, Universität Klagenfurt, Klagenfurt 9020, Austria
{pharmbrust,kerstinma}@edu.aau.at
[2] Lakeside Labs Gmbh, 9020 Klagenfurt, Austria
truden@lakeside-labs.at

Abstract. Attended home delivery services like online grocery shopping services require the attendance of the customers during the delivery. Therefore, the Vehicle Routing Problem with Time Windows occurs, which aims to find an optimal schedule for a fleet of vehicles to deliver goods to customers. In this work, we propose three sweep algorithms, which account for the family of cluster-first, route-second methods, to solve the Vehicle Routing Problem with Time Windows. In the first step, the customers are split into subsets such that each set contains as many as possible customers that can be served within one tour, e.g., supplied with one vehicle. The second step computes optimal tours for all assigned clusters. In our application, the time windows follow no special structure, and hence, may overlap or include each other. Further, time windows of different lengths occur. This gives additional freedom to the company during the planning process, and hence, allows to offer discounted delivery rates to customers who tolerate longer delivery time windows. Our sweep algorithms differ in the clustering step. We suggest a variant based on the standard sweep algorithm and two variants focusing on time window length and capacity of vehicles. In the routing step, a Mixed-Integer Linear Program is utilized to obtain the optimal solution for each cluster. The paper is concluded by a computational study that compares the performance of the three variants. It shows that our approach can handle 1000 customers within a reasonable amount of time.

Keywords: Vehicle routing · Time windows · Sweep algorithm · Attended home delivery · Transportation · Logistics

1 Introduction

The popularity of Attended Home Delivery (AHD) services, e.g., online grocery shopping services, increased within the last years. Especially, due to the current Covid-19 pandemic, this trend is continuing. For example in Western Europe, see [11], the share of online buyers is predicted to increase from 67% in 2020 to 75% in 2025. Therefore, the online share of groceries changed from 3.4% in 2019 to 5.3% in 2020 and is expected to reach 12.6% by 2025.

All AHD services have in common that the customer must attend the delivery of the goods or the provision of the booked service. To manage this in an effective manner,

B. Dorronsoro et al. (Eds.): OLA 2022, CCIS 1684, pp. 135–144, 2022.
https://doi.org/10.1007/978-3-031-22039-5_11

customers can typically choose among several time windows during which he or she is available to receive the ordered goods or to supervise the service provision. We consider an application where different time window lengths occur. This allows an operator of a delivery service to offer discounted delivery rates to customers who tolerate longer delivery time windows as such offers are easier to include within a tour. Typically, the company aims to minimize the overall delivery costs and therefore, a variant of the Vehicle Routing Problem (VRP), the so-called Vehicle Routing Problem with Time Windows (VRPTW), occurs. A comprehensive introduction to the VRP can be found in [13]. The authors give an overview of different VRP types with the help of applications, case studies, heuristics, and integer programming approaches. Another overview and classification of recent literature considering varieties of the Vehicle Routing Problem can be found in [2]. For an overview of exact methods for the VRPTW we refer to [1]. Heuristic methods are reviewed in [3] and [4], and a compact review of exact and heuristic solving approaches for the VRPTW can be found in [6].

Due to the fact that in AHD services mostly a large number of customers are delivered, solving this problem to optimality within reasonable time is rarely possible. In practice, heuristics are applied to produce delivery schedules of high quality within a short amount of time. A common approach is the *cluster-first, route-second* method. In the first step, the customers are split into subsets, so-called clusters, such that each set contains as many as possible customers that can be served within one tour, e.g., supplied with one vehicle. In the second step, an optimal tour for each cluster is computed. [7] originally introduced the *Sweep Algorithm* for the VRP. Using the analogy of clock hands, the depot is placed at the center of the plane. A clock hand then sweeps across the plane. The angle of the hand increases while a customer is inserted in the current cluster, if a feasible insertion is possible, or otherwise within a new cluster. A recent study, see [5], applies the Sweep Algorithm for a VRP occurring in an install and maintenance service for smart meter devices. In [12], the sweep algorithm only considers time windows when computing the routes for each vehicle but not during the clustering step. [8] consider time windows already in the clustering step. However, they make use of a special time window structure, where the time windows are non-overlapping.

In this work, we propose the following three variants of performing the clustering step of a sweep algorithm for a VRPTW:

- a variant based on the standard sweep algorithm [7],
- a variant that takes the length of the delivery time windows into account, and
- a variant considering both the lengths of the delivery time windows and the vehicles' capacities.

We consider time windows with no special structure, and hence, they can overlap or contain each other. A Mixed-Integer Linear Program (MILP), which is stated in [9], is applied for deciding the feasibility of a cluster of customers, and for obtaining the optimal tour for each cluster. Considering non-overlapping time-windows, like in [8], would lead to a more efficient MILP formulation (see [9] for a comparison of both MILP approaches). However, most modern routing applications use time windows, which can overlap each other.

For the computational study, we use a large variety of benchmark instances that have been carefully constructed such that they resemble real-world data. We consider different capacities of the vehicles and therefore, the tour lengths can differ. This study shows that our approach is capable of finding good initial solutions for instances containing up to 1000 customers within a short amount of time.

2 Mathematical Formulation

Now let us introduce the notation required to define the VRPTW. An instance of the VRPTW is defined by:

- A set of time windows $\mathscr{W} = \{w_1, \ldots, w_q\}$, where each $u \in \mathscr{W}$ is defined through its start time s_u and its end time e_u with $s_u < e_u$.
- A set of customers $\mathscr{C}$, $|\mathscr{C}| = n$, with corresponding order weight function $c\colon \mathscr{C} \to]0, C]$, where $C \in \mathbb{R}_{>0}$ denotes the given vehicle capacity, and a service time function $s\colon \mathscr{C} \to \mathbb{R}_{>0}$.
- A function $w\colon \mathscr{C} \to \mathscr{W}$ that assigns a time window to each customer during which the delivery vehicle must arrive.
- A depot d from which all vehicles depart from and return to, $\overline{\mathscr{C}} := \mathscr{C} \cup \{d\}$.
- A travel time function $t\colon \overline{\mathscr{C}} \times \overline{\mathscr{C}} \to \mathbb{R}_{\geq 0}$.

In the following, we state some basic definitions and we define the following notation $[u] := [1, \ldots, u]$, where $u \in \mathbb{N}$.

A *tour* of n customers consists of a set $\mathscr{A} = \{a_1, a_2, \ldots, a_n\}$ and the indices of the costumers refer to the order of the customers. Each customer in the tour has a corresponding arrival time α_{a_i}, $i \in [n]$, during which the vehicles are scheduled to arrive. Furthermore, each tour has assigned *start* and *end times* that we denote as $start_{\mathscr{A}}$ and $end_{\mathscr{A}}$, respectively. Hence, the vehicle executing tour $\mathscr{A}$ can leave from the depot d no earlier than $start_{\mathscr{A}}$ and must return to the depot no later than $end_{\mathscr{A}}$.

We denote a tour $\mathscr{A}$ as *capacity-feasible*, if $\sum_{i=1}^{n} c(a_i) \leq C_{\mathscr{A}}$. The special case that the capacity of a single customer exceeds the capacity limit of the vehicles cannot occur.

If for each customer of a tour holds that the delivery of the goods occurs within the time window $u = w(a_i)$, i.e., $s_{w(a_i)} \leq \alpha_{a_i} \leq e_{w(u_i)}$, and there is enough time for fulfilling the order and traveling to the next customer, i.e., $\alpha_{a_{i+1}} - \alpha_{a_i} \geq s(a_i) + t(a_i, a_{i+1})$, for all $i \in [n]$, then we call it *time-feasible*.

A *feasible* tour is capacity- and time-feasible and a schedule $\mathscr{S} = \{\mathscr{A}, \mathscr{B}, \ldots\}$ consists of feasible tours where each customer occurs exactly once.

Each element in $\overline{\mathscr{C}}$ has geographical coordinates in the two-dimensional plane and we assume that the travel times are correlated to the geographical distances. In general, the travel times are somehow related to, but not completely determined by the geographical distances. In our application, we typically deal with asymmetric travel time functions, for which the triangle inequalities, i.e., $t(a, c) \leq t(a, b) + t(b, c)$, $a, b, c \in \overline{\mathscr{C}}$, do not hold. In general, $t(a, b) = t(b, a)$, where $a, b \in \overline{\mathscr{C}}$, is not guaranteed for an asymmetric travel time function.

The length of a time window $u \in \mathscr{W}$ is defined as $e_u - s_u$. Each customer can choose among time windows of length 10, 20, 30, 60, 120, or 240 min.

Two time windows u and v are overlapping, if $u \cap v \neq \emptyset$, and non-overlapping, if $u \cap v = \emptyset$. Time window u contains time window v, if $s_u \leq s_v$, and $e_v \leq e_u$. If time window u contains time window v, then these time windows overlap each other. It must not hold that a time window s which overlaps t, also contains t.

In this work we consider three objectives, namely $\lambda_1, \lambda_2, \lambda_3$:

- The first one is the *number of vehicles used*, i.e., $\lambda_1(\mathscr{S}) := |\mathscr{S}|$.
- Secondly, the sum of all *tour durations* is denoted by the *schedule duration* $\lambda_2(\mathscr{S})$, i.e., $\sum_{\mathscr{A} \in \mathscr{S}} \lambda_2(\mathscr{A})$, where $\lambda_2(\mathscr{A}) := t(d, a_1) + \alpha_{a_n} - \alpha_{a_1} + s(a_n) + t(a_n, d)$.
- Thirdly, the sum of all *tour travel times* is denoted by the *schedule travel time* $\lambda_3(\mathscr{S})$, i.e., $\sum_{\mathscr{A} \in \mathscr{S}} \lambda_3(\mathscr{A})$, where $\lambda_3(\mathscr{A}) := t(d, a_1) + \sum_{i=1}^{n-1} t(a_i, a_{i+1}) + t(a_n, d)$.

Similar to [12] and [8], we aim to minimize these three objectives with respect to the lexicographical order $(\lambda_1, \lambda_2, \lambda_3)$, since providing a vehicle is usually the most expensive cost component, followed by the drivers' salaries, and the costs for fuel.

3   Sweep Strategy

A sweep algorithm is based on the polar coordinate angles of the customers $\mathscr{C}$, where the depot d lies in the center of the grid and $\theta(a) \in [0, 2\pi[$ denotes the angle component of a customer $a \in \mathscr{C}$. We choose the zero angle, i.e., the starting angle, according to [8] such that it splits the two neighboring customers with the largest angle gap, i.e., $\max_{a,b \in \mathscr{C}} \theta(a) - \theta(b)$.

We propose the following general strategy to obtain solutions for a given VRPTW instance consisting of the following steps: 1. Apply one of the methods that obtain a feasible clustering of $\mathscr{C}$. 2. Compute the optimal route for each cluster.

In both steps, the *Traveling Salesperson Problem with Time Windows* (TSPTW) occurs as a subproblem to check time-feasibility or to obtain the optimal solution of a single tour. Next, we give further information about the two steps.

Clustering: We apply the following variants of clustering algorithms in Step 1.

- Traditional Sweep
- Sweep Algorithm depending on Time Window Length
- Sweep Algorithm depending on Time Window Length and Overall Capacity

Each of the three algorithms is described in more detail below. Moreover, we illustrate each variant with a toy example. The depot is located in the center of the coordinate system and the capacity of each vehicle is 10. In the visualizations we choose the direction of the zero angle $\theta_0 = 90°$ as three o'clock and increase the angle in each step counterclockwise. Depending on the selected sweep algorithm we check in each iteration the capacity-feasibility or time- and capacity-feasibility. Note that the MILP is only used if the tour is capacity-feasible and the current tour candidate is not time-feasible, i.e., cannot feasible inserted without changing the order of the already placed customers. Then, the exact approach tries to find any time-feasible tour including the current costumers.

After obtaining an initial clustering using one of the heuristics, we determine efficient tours for each vehicle.

Routing: In Step 2, a tour for each cluster is obtained by solving the TSPTW-MILP with lexicographical objective (λ_2, λ_3) to optimality. We apply an MILP formulation that has been proposed in [9] for solving the TSPTW. Following the lexicographical order, the MILP first minimizes the tour duration $\lambda_2(\mathscr{A})$, and then the tour travel time $\lambda_3(\mathscr{A})$ while keeping $\lambda_2(\mathscr{A})$ fixed.

In the following subsections, we describe the different clustering steps in more detail.

3.1 Traditional Sweep

The first variant reflects the traditional algorithm by [7] except checking for both, time- and capacity-feasibility. It works as follows:

1. Compute the zero angle θ_0 and sort the customers according to their polar coordinate angles starting with θ_0.
2. Start with an empty cluster.
3. We check, if the next customer from the sorted set can be feasibly inserted within the current cluster.
   - If capacity- and time-feasibility holds, we add the new customer to the current cluster.
   - Otherwise, we initiate a new cluster with the current customer.
4. We repeat step 3 until all given customers are assigned within a cluster.

Therefore, the result of the traditional sweep algorithm is a set of capacity- and time-feasible clusters.

A simple example of the traditional sweep algorithm with five customers is depicted in Fig. 1. The weights of the customers are given in their boxes, the capacity of each vehicle is 10, a depot in the center and we start with the zero angle at three o'clock (90°). In each step, we increase the angle counterclockwise, such that a new customer is added to the current cluster. Then we check the capacity- and time-feasibility of it and if necessary, we increase the number of clusters.

3.2 Sweep Algorithm Depending on Time Window Length

In this subsection, we describe a sweep algorithm that takes the lengths of the delivery time windows into account. First, we sort all customers in ascending order of their time window lengths. Due to the high number of customers there are many customers with the same time window length and therefore, we start with the following procedure:

- The customer with the lowest time window length is added to the first cluster. If there is more than one customer with the same length, the sweep algorithm increases the angle beginning from the zero angle and selects the first customer of the set.
- In each sweep iteration: We increase the angle and consider the next customer within the current time window length and try to add this customer to the current cluster:
  - If the tour is time- and capacity-feasible, add the customer to the current cluster.
  - Else, increase the number of clusters and add the customer to the next cluster.

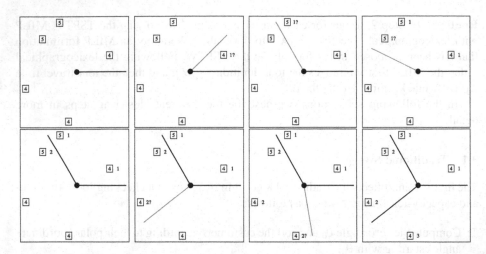

Fig. 1. Example of the traditional sweep method with one depot in the center. The capacity of each vehicle is 10 units, the weight of each customer is given in their boxes, and the number on the right side of the boxes denotes the number of the scheduled cluster.

- If no customer remains, increase the current time window length and start with the sweep iteration again, until all customers are scheduled within a cluster.

For further clarifying this algorithm, we consider a toy example with the customers given in Table 1.

Table 1. Toy example of customers sorted by their time window length for sweep algorithm depending on time window length.

	Required capacity	Time window properties		
		Start time	End time	Length
Customer 1	5	09:00	10:00	60
Customer 2	4	14:00	15:00	60
Customer 3	4	09:30	11:30	120
Customer 4	5	12:00	14:00	120

In our example, the capacity of each vehicle is 10. We start with the lowest time window length. There are two customers (customer 1 and customer 2) with the same time window length. Due to the given angles, customer 1 is added to the first cluster. In the following step, we consider customer 2, which we try to add to the current cluster, and therefore we check, if the tour remains time- and capacity-feasible after an insertion. In this case, the current tour satisfies both conditions, and hence, we add customer 2 to the first cluster. Next, we consider customer 3, which cannot be added to the current cluster, due to the capacity limit of the vehicle. Therefore, we increase the number of clusters and add customer 3 to the second cluster. Then we try to add the last customer in the toy example, and hence, the time- and capacity-feasibility of the second tour with the next customer is checked. The insertion is possible, and hence, customer 4 is added to the current cluster. Our algorithm terminates with two clusters consisting of two customers each.

3.3 Sweep Algorithm Depending on Time Window Length and Overall Capacity

Again we sort the customers by their time window lengths, and afterward by their polar coordinate angles. In contrast to Subsect. 3.2, we now reserve half of the vehicle capacity for customers with longer time window lengths. The remaining procedure stays the same. As a result, this algorithm returns clusters, which are up to a half filled with customers having a time window length of 10, 20, or 30 min and the remaining capacity is filled with customers having a time window length of 60, 120, or 240 min. The idea behind this allocation is to ensure that the created tours have a proper mixture of short and long time windows. This property shall increase the robustness of the tours against delays occurring during the delivery process. The guaranteed portion of long delivery time windows introduces some slack such that late deliveries become less likely.

Now, we continue with a toy example and all information about the customers are given in Table 2.

Table 2. Toy example of customers sorted by their time window length for sweep algorithm depending on time window length and overall capacity constraints.

	Required capacity	Time window properties		
		Start time	End time	Length
Customer 1	5	09:00	09:10	10
Customer 2	4	14:00	14:30	30
Customer 3	5	15:00	15:30	30
Customer 4	4	09:30	10:30	60
Customer 5	4	11:30	13:30	120
Customer 6	5	12:00	14:00	120

First, we add the first customer (with the shortest time window length) to the first cluster. As we reserve half of the capacity for customers with time windows longer than 30 min, customer 3 (selected due to the polar coordinate angles of customer 2 and customer 3) is added to a new cluster, and the same procedure applies to customer 2. According to the polar coordinate angles, we try to add customer 5 to the first tour, and therefore, we check, if the cluster is time- and capacity-feasible. Both constraints are satisfied, and customer 5 is added to the first cluster. Next, we further increase the angle of the hand, and we consider customer 6 to be next. We notice that there is not sufficient capacity left in the first cluster. Hence, we assign customer 6 to the second cluster after checking the constraints. Finally, customer 4 can be feasibly inserted into the third cluster. Therefore, the algorithm returns three clusters containing two customers each.

4   Computational Results

In this work, we introduce a benchmark set that imitates urban settlement structures. Customers can choose from time windows of different lengths, i.e., 10, 20, 30, 60, 120, or 240 min. The operation time of the vehicles is ten hours a day and the time windows are sampled randomly between the operation times. The customers are distributed on a 20 km × 20 km square grid that is roughly of the same size as the city of Vienna, Austria. Furthermore, 80% of the customers are arranged within randomly selected clusters and only 20% are uniformly distributed on the grid. The Euclidean distance is used to calculate the distance between two customers. We assume an average travel speed of 20 km/h (as proposed by [10]) to calculate the travel times.

The service time at each customer is set to five minutes and the order weights are sampled from a truncated normal distribution centered around five units. In the computational study, we consider different numbers of customers, namely $|\mathscr{C}| = \{250, 500, 750, 1000\}$, and the capacity of the vehicles is set to $C_{\mathscr{A}} = 200$, $\mathscr{A} \in \mathscr{S}$. The instances are available at http://dx.doi.org/10.13140/RG.2.2.20934.60480. We use an Ubuntu Mint 20 machine equipped with an Intel Xeon $E5 - 2630V3@2.4$ GHz 8 core processor and 132 GB RAM and Gurobi 8.1.1 in single-thread mode to solve the Mixed-Integer Linear Programs. For each instance, we apply our three methods and the average results over 10 instances for the different numbers of customers are given in Tables 3, 4, 5 and 6. We observe similar behavior for all instance sizes considered. The Traditional Sweep is much faster compared to the two other methods for both steps in the sweep algorithm. As the first method obtains up to four times the number of clusters, the number of customers within a tour is very low, and therefore, the routing step is very efficient using an MILP approach. However, the number of vehicles is crucial, and thus, the methods two and three perform in a costefficient manner. There is no big difference in the results between the two methods with and without considering the overall capacity. However, it is expected that the third sweep variant produces more robust tours as described in Subsect. 3.3. Thus, we are able to improve robustness without any loss of quality with respect to all three objectives. Further, the traditional sweep badly performs with respect to λ_2 but slightly improves λ_1 compared to the other two variants. We are able to find clusters for all algorithms within eight minutes for up to 1000 customers. The runtime for the routing step is up to one and a half hours for variants two and three.

Table 3. Average results with 250 customers over 10 instances each. We denote the runtime by t. λ_1 depicts the number of vehicles used, λ_2 gives the schedule duration, and λ_3 denotes the total travel time.

Runtime/Objectives	Clustering		Routing		
	t	λ_1	t	λ_2	λ_3
Units	[m:ss]		[mm:ss]	[hhh]	[hh]
Traditional Sweep	0:11	32.7	00:01	222	50
Time window length	1:29	11.7	14:56	46	67
Length & overall capacity	1:36	11.9	14:29	46	68

Table 4. Average results with 500 customers over 10 instances each. We denote the runtime by t. λ_1 depicts the number of vehicles used, λ_2 gives the schedule duration, and λ_3 denotes the total travel time.

Runtime/Objectives	Clustering		Routing		
	t	λ_1	t	λ_2	λ_3
Units	[m:ss]		[mm:ss]	[h]	[h]
Traditional Sweep	0:46	65.9	00:01	460	101
Time window length	3:22	19.8	33:57	76	116
Length & overall capacity	3:48	20.0	40:27	77	116

Table 5. Average results with 750 customers over 10 instances each. We denote the runtime by t. λ_1 depicts the number of vehicles used, λ_2 gives the schedule duration, and λ_3 denotes the total travel time.

Runtime/Objectives	Clustering		Routing		
	t	λ_1	t	λ_2	λ_3
Units	[m:ss]		[h:mm:ss]	[h]	[h]
Traditional Sweep	1:51	101,3	0:00:02	714	144
Time window length	6:27	29,4	0:57:51	111	165
Length & overall capacity	6:31	29,8	1:00:40	114	155

Table 6. Average results with 1000 customers over 10 instances each. We denote the runtime by t. λ_1 depicts the number of vehicles used, λ_2 gives the schedule duration, and λ_3 denotes the total travel time.

Runtime/Objectives	Clustering		Routing		
	t	λ_1	t	λ_2	λ_3
Units	[m:ss]		[h:mm:ss]	[h]	[h]
Traditional Sweep	1:53	149.6	0:00:03	1040	196
Time window length	7:26	37.1	1:16:02	151	203
Length & overall capacity	7:49	36.9	1:23:27	152	200

5 Conclusion

In this paper, we considered several sweep algorithms, which belong to cluster-first, route-second methods, for the Vehicle Routing Problem with Time Windows. The first step of the algorithm clusters the customers according to their polar coordinate angles originating from the depot as the center point of the grid. Secondly, the tour for each cluster is determined. Considering the clustering step, we introduced a variant based on the standard sweep algorithm and two variants focusing on time window length and capacity of vehicles. Further, a benchmark set with different customer sizes is provided. Each customer chooses the length of the time window within a given set, namely 10, 20, 30, 60, 120, and 240 min. Our computational study showed that the heuristic is able to cluster 1000 customers within eight minutes. In the routing step, the optimal tours are calculated by a Mixed-Integer Linear Program, which results in runtimes of up to one and a half hours. It remains for future work to apply a heuristic approach that gathers good quality solutions in a fraction of time.

Acknowledgments. This work was supported by Lakeside Labs GmbH, Klagenfurt, Austria, and funding from the European Regional Development Fund and the Carinthian Economic Promotion Fund (KWF) under grant 20214/31942/45906.

References

1. Baldacci, R., Mingozzi, A., Roberti, R.: Recent exact algorithms for solving the vehicle routing problem under capacity and time window constraints. Eur. J. Oper. Res. **218**(1), 1–6 (2012)
2. Braekers, K., Ramaekers, K., Nieuwenhuyse, I.V.: The vehicle routing problem: state of the art classification and review. Comput. Ind. Eng. **99**, 300–313 (2016)
3. Bräysy, O., Gendreau, M.: Vehicle routing problem with time windows, Part I: route construction and local search algorithms. Transp. Sci. **39**(1), 104–118 (2005)
4. Bräysy, O., Gendreau, M.: Vehicle routing problem with time windows, Part II: metaheuristics. Transp. Sci. **39**(1), 119–139 (2005)
5. Bucur, P.A., Hungerländer, P., Jellen, A., Maier, K., Pachatz, V.: Shift planning for smart meter service operators. In: Haber, P., Lampoltshammer, T., Mayr, M., Plankensteiner, K. (eds.) Data Science – Analytics and Applications, pp. 8–10. Springer, Wiesbaden (2021). https://doi.org/10.1007/978-3-658-32182-6_2
6. El-Sherbeny, N.A.: Vehicle routing with time windows: an overview of exact heuristic and metaheuristic methods. J. King Saud Univ. **22**, 123–131 (2010)
7. Gillett, B.E., Miller, L.R.: A heuristic algorithm for the vehicle-dispatch problem. Oper. Res. **22**(2), 340–349 (1974)
8. Hertrich, C., Hungerländer, P., Truden, C.: Sweep algorithms for the capacitated vehicle routing problem with structured time windows. In: Fortz, B., Labbé, M. (eds.) Operations Research Proceedings 2018. ORP, pp. 127–133. Springer, Cham (2019). https://doi.org/10.1007/978-3-030-18500-8_17
9. Hungerländer, P., Truden, C.: Efficient and easy-to-implement mixed-integer linear programs for the traveling salesperson problem with time windows. Transp. Res. Procedia **30**, 157–166 (2018)
10. Pan, S., Giannikas, V., Han, Y., Grover-Silva, E., Qiao, B.: Using customer-related data to enhance e-grocery home delivery. Ind. Manag. Data Syst. **117**(9), 1917–1933 (2017)
11. Savills: European Food and Groceries Sector, European Commercial, Savills Commercial Research (2021). https://pdf.euro.savills.co.uk/european/europe-retail-markets/spotlight---european-food-and-groceries-sector---2021.pdf. Accessed 19 Dec 2021
12. Solomon, M.M.: Algorithms for the vehicle routing and scheduling problems with time window constraints. Oper. Res. **35**(2), 254–265 (1987)
13. Toth, P., Vigo, D. (eds.): The Vehicle Routing Problem: Problems, Methods, and Applications, 2nd edn. Society for Industrial and Applied Mathematics, Philadelphia (2014)

Improving the Accuracy of Vehicle Routing Problem Approximation Using the Formula for the Average Distance Between a Point and a Rectangular Area

Daisuke Hasegawa[1]([✉]) [iD], Yudai Honma[1] [iD], Naoshi Shiono[2] [iD], and Souma Toki[3] [iD]

[1] The University of Tokyo, Bunkyo, Tokyo, Japan
hasega60@e.u-tokyo.ac.jp, yudai@iis.u-tokyo.ac.jp
[2] Kanagawa Institute of Technology, Atsugi, Kanagawa, Japan
na-shiono@ic.kanagawa-it.ac.jp
[3] Tokyo Gas Co., Ltd., Minato, Tokyo, Japan
toki.s@tokyo-gas.co.jp

Abstract. The continuous approximation model of VRP analyzes cost at the planning and strategic analysis stage in the delivery and logistic field. Most previous studies used the tour distance between customers and the linehaul distance between the depot and the customers. This study focused on the linehaul distance, and we applied the formula for the average length in a right triangle and presented the average distance between depot and service area. Our proposed model is tested in instances with different locations of the depot and the service area, trucks, and customers. Regression results indicate that the approximation model improves accuracy if the depot locates outside the service area. Our results can be applied when planning deliveries, for cases where the depot is located at the edge of the city and outside the delivery area, and planning area segmentation.

Keywords: Vehicle routing problem · Distance estimation · Continuous approximation approach

1 Introduction

The travel salesman problem (TSP) and vehicle routing problem (VRP) are methods for minimizing the cost of transportation using single or multiple vehicles from one location to a customer. This problem is extremely important not only in the logistics sector but also in the public transportation sector, where new services such as ride-sharing and car-pooling are being developed. With the improvement in computers and algorithms, the number of possible optimal solutions has increased; however, there are many cases where an approximate solution is required. For example, if the number of demand points is too high for an optimal solution, the number of vehicles will be too high, or the specific location of the demand points will be unknown. Some mathematical models (commonly called continuous approximation models) can be used to approximate

B. Dorronsoro et al. (Eds.): OLA 2022, CCIS 1684, pp. 145–156, 2022.
https://doi.org/10.1007/978-3-031-22039-5_12

the optimal solution. Currently, numerous studies have been conducted on continuous approximation models for the TSP and VRP. Some of these models include the average distance between the depot and the delivery area, considering the case where the depot, which is a base for vehicles, is not at the center of the delivery area. Many previous studies have used the linehaul distance for such cases. However, they used simplifying assumptions and did not strictly reflect the shape and location of a region. This study aims to improve the approximation accuracy of the continuous approximation model using an analytically derived value of the average distance between a depot and the delivery service area.

The rest of this paper is organized as follows. Section 2 provides a literature review of the VRP and TSP length calculations obtained using the continuous approximation approach. Section 3 presents the approximation result of the average distance between points and a rectangular area and its application to the continuous approximation model of the VRP. Section 4 describes the experimental design and estimation results. Finally, Sect. 5 concludes the paper.

2 Literature Review

Most of the continuous approximation models for the TSP and VRP have been based on the study conducted by Beardwood et al. [1], who proved the result generally known as the BHH formula. For a set of n random points in a i-dimensional space $\mathbb{R}^i$, the length of an optimal tour through n points D^* satisfies:

$$\lim_{n \to \infty} \frac{D^*}{n^{(i-1)/i}} = k_i i^{\frac{1}{2}} [v(\Psi)]^{\frac{1}{i}}, \tag{1}$$

where the measure of the Lebesgue-measurable set Ψ is denoted by $v(\Psi)$, and k_i is a constant that depends on i. This formula is quite complicated, and Eilon et al. [6] showed a simple explanation of the planar case ($i = 2$), where S is planar area of and k is used instead of k_2:

$$\frac{D^*}{\sqrt{n}} \to k\sqrt{S} \quad \text{if } n \to \infty, \tag{2}$$

and if $n \in \mathbb{N}$, Eq. (2) can be rewritten as:

$$D^* \approx k\sqrt{nS}. \tag{3}$$

The value of k is an unknown constant; nevertheless, it has been estimated in several previous studies. In the case of the Manhattan distance metric, Jaillet [11] estimated $k = 0.97$, and Stein [18] estimated $k = 0.765$ using the Euclidean distance metric. Cook et al. [3] showed that k is correlated to n and estimated it to be $0.625 \leq k \leq 0.920$ for n values between 100 and 2000.

In addition, the BHH formula can be extended to a VRP that has capacitated vehicles based on the depot visit customers in the area. Daganzo [4] proposed a simple formula for the optimal length of VRP D_m^* with m routes:

$$D_m^* = k\sqrt{nS} + 2Rm, \tag{4}$$

where the value of m is obtained from the number of customers n and the maximum capacity of vehicle C ($m = n/C$); parameter $k = 0.57$, and R is the distance from the depot to a random point in the area. In this formula, the first term refers to the tour length estimated by the BHH formula, and the second term is generally called the linehaul distance, which is the distance from the depot to the customer. The linehaul distance contributes to the estimation when the depot is not located at the center of the service area. Some studies have developed regression tests to estimate the TSP optimal length D^* and VRP optimal length D_m^*. Chien [2] suggested this approximation for a depot located at the corner of the area. Let B be the size of the boundary box enclosing all points; in this case, the total distance in the case of one truck D_1^* (TSP route length) can be expressed as:

$$D_1^* = 0.67\sqrt{nB} + 2.1R. \tag{5}$$

Figliozzi [7] presented the following as a model of the VRP containing the linehaul distance:

$$D_m^* = k\left(\frac{n-m}{n}\right)\sqrt{nS} + 2Rm \tag{6}$$

where the parameter k is estimated to be $0.62 \leq k \leq 0.90$, and average $k = 0.77$ by linear regression, with a high accuracy for approximating D_m^*.

Most previous studies have focused on the tour length, indicated by the first term, to improve the accuracy. Therefore, the assumption of the linehaul distance R is simplified. In Eq. (4) proposed by Daganzo [4], R is determined to check whether the depot o is located in service area $\mathcal{D}$ as follows:

$$R = \begin{cases} \left(\sqrt{S}/6\right)\left(\sqrt{2} + \log\tan(3\pi/8)\right) \cong 0.382\sqrt{S}, & (o \in \mathcal{D}) \\ \hat{d} & (o \notin \mathcal{D}). \end{cases} \tag{7}$$

where $\hat{d}$ is the distance from the depot o to the center of gravity of the service area $\mathcal{D}$. Also, Franceschetti et al. [8] used the closest point in the service area. Huang et al. [9] used the expected distance to reach the first point by taking the square root of the sum of the square of the expected longitudinal distance and the square of the expected transverse distance.

However, the tour start point exists randomly in the service area, and we expect that calculating the average distance by considering the area and shape of the service area can help improve the estimation accuracy of D_m^*. For this purpose, we tested using two different linehaul distances for variable R in Daganzo's model (4) and Figlilozzi's model (6). One is the distance from the center of gravity of the area used in previous studies, and the other is the average distance between the point and the service area calculated using a probabilistic approach.

3 Approximation of the Average Distance Between a Point and an Area

In this section, we first describe the problem setting. Next, we present the methods for determining the distance between points and an area.

3.1 Problem Setting

In the continuous approximation approach for TSP and VRP reported in previous studies, the service area was simplified to a rectangular shape [4, 8, 9], circular/elliptical shape [16], and a ring-radial network [10, 15]. However, the shape of the area does not significantly affect the quality of the approximation [5, 13]. Thus, we model the service area as a rectangular shape for ease of numerical experiments and assume that the depot position is fixed and that the service area can be flexibly changed to account for the differences in the linehaul distance. We assume a rectangular service area $\mathcal{D}$ with horizontal length A and vertical length B. The n number of customers in $\mathcal{D}$ are distributed in the form of a continuous uniform distribution and are picked up by m trucks based at depot o (Fig. 1). We calculate the average distance of a right triangle area using the model proposed by Koshizuka and Kurita [12], Kurita [14] and apply it to the rectangular regions.

3.2 Model

Let the right triangle δ, shown in Fig. 2, be a depot o as an acute vertex. α is the edge from o to the right angle a, and β is the other edge from o to b. The probability density function $\varphi(x)$ of the distance from o to a point x in δ can be expressed as:

$$\varphi(x) = \frac{L(x)}{S} \tag{8}$$

$$S = \frac{\alpha}{2}\sqrt{\beta^2 - \alpha^2} \tag{9}$$

Here, S is the size of δ, and $L(x)$ is the perimeter of the fan shapes with x radius and is obtained from cases shown in Figs. 2(i) and (ii):

$$L(x) = \begin{cases} \text{(i) } x \arccos\dfrac{\alpha}{\beta} & (0 \leq x \leq \alpha) \\[2mm] \text{(ii) } x\left(\arccos\dfrac{\alpha}{\beta} - \arccos\dfrac{\alpha}{x}\right) & (\alpha < x \leq \beta) \end{cases} \tag{10}$$

where $\arccos \alpha/\beta$ is the angle θ between α and β. Thus, the average distance d_δ^* can be expressed as:

$$d_{oab}^* = \int_0^\beta x\varphi(x)\mathrm{d}x \tag{11}$$

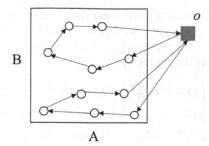

Fig. 1. Service area and depot.

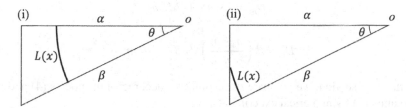

Fig. 2. Location of service area depot.

Equation (11) can be reorganized as follows:

$$d^*_{oab} = \frac{1}{3}\left(\beta + \frac{\alpha^2}{\sqrt{\beta^2 - \alpha^2}}\, \ln\frac{\beta + \sqrt{\beta^2 - \alpha^2}}{\alpha^2}\right) \tag{12}$$

Subsequently, we can calculate the average distance to the rectangular area by combining the right triangles. There are four cases (I to IV in Fig. 3) depending on whether the location of o is within the horizontal or vertical extent of the area. Figure 4 shows the calculation method for the average distance between Cases I and IV.

For a rectangular area $\mathcal{D}_{abcd}$, the average distance d^* from o to $\mathcal{D}_{abcd}$ can be expressed as:

$$d^* = \frac{LD_{abcd}}{S_{abcd}} \tag{13}$$

where S_{abcd} is the size of $\mathcal{D}_{abcd}$, and the total distance LD_{abcd} is calculated by combining the LD of the rectangles with o as its vertex. The intersection points of each edge and perpendicular line from depot o are $e, f, g,$ and h. The total distance LD_{aego} from o to the rectangular area $\mathcal{D}_{aego}$ (upper left in Case I) with o as its vertex can be expressed as:

$$LD_{aego} = S_{aeo}d^*_{aeo} + S_{ago}d^*_{ago} \tag{14}$$

where the size of the right triangle is denoted by S_{aeo}. In Case I, where o is in $\mathcal{D}_{abcd}$, LD_{abcd} is obtained by summing the LD of the rectangles around o. In Cases II, III, and IV,

e, f, g, and h are the intersections of the line extending each edge and perpendicular line, and we removed the unnecessary part (the gray area in Fig. 4 (IV)) from the rectangle with o as its vertex that is greater than $\mathcal{D}_{abcd}$. Thus, let the coordinates of a be (a_1, a_2), and LD_{abcd} can be expressed as in Eq. (15).

$$LD_{abcd} = \begin{cases} \text{(I)} & LD_{aego} + LD_{ebof} + LD_{goch} + LD_{ofdh}, \\ \text{(II)} & LD_{ofgd} + LD_{ebof} - \left(LD_{ohgc} + LD_{eaoh}\right) \\ \text{(III)} & LD_{aeho} + LD_{ebof} - \left(LD_{cgho} + LD_{gdof}\right) \\ \text{(IV)} & LD_{ebof} - \left(LD_{eaoh} + LD_{gdof} - LD_{gcof}\right) \end{cases} \tag{15}$$

We used d^* as the linehaul distance R for models (4) and (6):

$$D_m^* = k\sqrt{nS} + 2d^*m \tag{16}$$

$$D_m^* = k\left(\frac{n-m}{n}\right)\sqrt{nS} + 2d^*m, \tag{17}$$

In the next section, we compare the estimation accuracies of models (4) and (16), and (6) and (17) by numerical experiments.

4   Experimental Setting and Results

In this section, we first explain the numerical experimental setup. Next, we evaluate the accuracy of the estimates obtained using the continuous approximation model.

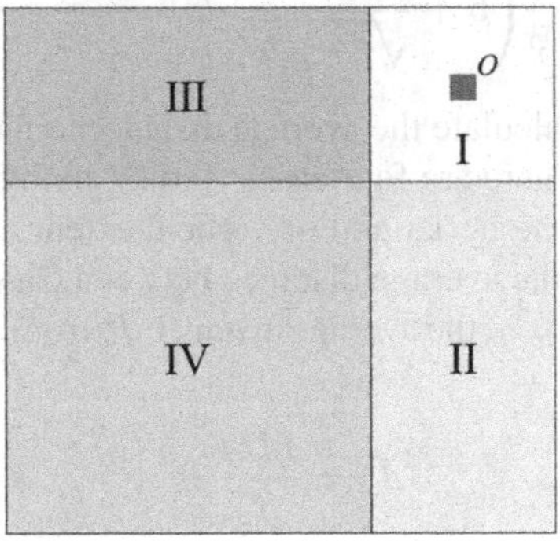

Fig. 3. Calculation condition for d^*.

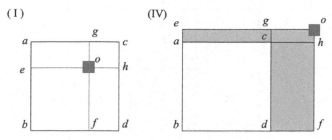

Fig. 4. Calculation method for d^* (Cases I and IV)

4.1 Experimental Setting

Numerical experiments were conducted to verify the accuracy of the continuous approximation model defined in Sect. 3. There are many instances of TSP and VRP, such as instance data from Solomon, TSPLib, and CVRPLib [17, 19–21]. For these examples, problems and solutions are available, including the spatial distribution of customers, vehicle capabilities, customer requirements, and customer time settings. However, in many instances, the depot location is included in the range of the demand distribution, and we cannot evaluate the difference in the linehaul distance, which is the subject of discussion in this study. Therefore, route optimizations with different area sizes and locations, demand patterns, and number of vehicles were obtained using mixed-integer linear programming (MILP).

Table 1. Experimental conditions.

Items		Values
Area settings	Area length (A, B)	(10000, 10000), (20000, 10000), (30000,10000)
	Depot location	(25000, 25000), (50000, 25000), (75000, 25000)
Layout patterns		9
Trucks		2, 3, 4, 5
Demands		10, 20, 30, 40, 50 points per trucks

Table 1 shows the details of the experimental conditions. Figure 5 shows the layout pattern of the depot o and service area $\mathcal{D}$. The area length is the length of the edges with $\mathcal{D}$ (Fig. 4, (A, B)); three types with different aspect ratios are assumed, and the position of o is changed on the basis of the aspect ratio of $\mathcal{D}$. The area layout has nine patterns (Fig. 5, Nos. 1 to 9) for the layout of o and $\mathcal{D}$. In the case of No. 9, o is located at the centroid of $\mathcal{D}$, which is the pattern with the shortest linehaul distance; conversely, in case No. 1, it is the pattern with the farthest distance from o. The number of trucks is assumed to range from 1 to 5, and the number of demands is assumed to range from 10 to 50 per truck, which is the same as the truck capacity.

We solved 540 instances (three area settings, nine layouts, four trucks, and five demands) 10 times by randomly changing the demand points within the area.

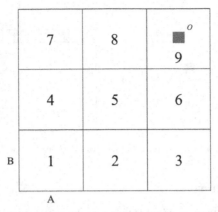

Fig. 5. Layout pattern of the depot and service area.

We used LocalSolver 9.5 to solve the VRP, and the CVRP algorithms for the problems that have been tested for performance by LocalSolver [22]. The computational tests were performed on two Windows 10 machines (3.5 GHz Intel Core i9 processor, with 64 GB RAM; 2.5 Xeon GHz processor with 64 GB RAM). The maximum running time was set to 300 s. However, for the pattern with the largest number of combinations (five trucks, demand 50 per truck), the optimality gap with the upper bound was less than 5%.

4.2 Results

To evaluate the prediction accuracy, the R-square value, root-mean-squared error (RMSE), and mean absolute percentage error (MAPE) were used. The RMSE and MAPE were calculated as follows:

$$RMSE = \sqrt{\frac{1}{n}\sum_i (y_i - f_i)^2} \tag{18}$$

$$MAPE = \frac{100}{n}\sum_i \left| \frac{(y_i - f_i)}{y_i} \right| \tag{19}$$

where the actual distance, which is calculated by MILP for instance i, is denoted y_i, and the estimated distance is denoted by f_i. The RMSE indicates the absolute error value for a specific distance. However, the longer the total distance, the greater the error. Therefore, the MAPE is used to determine the relative error.

Table 2 presents the fitting results for the linehaul distance and average distance. In the estimation, for model (4), we used $k = 0.57$, the value given in the original paper, and for model (6), we used the same value. In Figliozzi [7], k was fitted with real data [21], and the conditions of the numerical experiments were different from those in this study. For models (16) and (17), which are the focus of this study, parameter k was fitted using the maximum likelihood estimation method. In Table 2, all the models have good R^2 values. However, models (16) and (17), which use a continuous approximation term for the linehaul distance, have a better RMSE and MAPE performance than models (4) and (6).

Table 2. Model fit comparison.

Model	k	R^2	RMSE	MAPE
(4)	0.570	0.995	10087.0	3.4%
(16)	0.529	0.998	6830.2	2.5%
(6)	0.570	0.995	9944.8	3.5%
(17)	0.549	0.998	6883.5	2.5%

Table 3. Model fit comparison by number of trucks.

Model	RMSE				MAPE			
	Trucks				Trucks			
	2	3	4	5	2	3	4	5
(4)	7794.2	9350.5	10723.6	11992.3	3.9%	3.5%	3.1%	2.9%
(16)	6870.8	6627.1	6694.2	7118.0	3.6%	2.6%	2.0%	1.8%
(6)	8347.9	9500.4	10356.9	11330.7	4.4%	3.7%	3.1%	2.9%
(17)	6775.1	6688.2	6838.3	7220.4	3.7%	2.7%	2.1%	1.8%

Next, we compare the performance of each model based on the number of trucks and the layout of the area and depot to analyze the factors that can help improve the accuracy. Table 3 shows the comparison results of the number of trucks. The number of trucks is proportional to the linehaul distance, as shown in the second term of each model. Therefore, the greater the number of trucks, the better the RMSE and MAPE performance. However, the difference is highest when the number of vehicles is four, and above that, the performance tends to saturate.

Figure 6 shows a comparison of the RMSE and MAPE values of models (4) and (16) for different area layouts, as shown in Fig. 5. The performance of No. 1 to 8, which are the area layouts away from the depot, is improved. In particular, No. 6 exhibits the best performance. Figure 7 shows the reason for this. This figure shows the difference in the distance from the depot to the center of gravity $\hat{d}$ or rectangular areas d^* when the depot is located at (75000, 25000), and the area edges $A = 30000$ and $B = 10000$ (thus, the aspect ratio is 3.0) are moved. When the area does not include the depot (in the red frame), a large difference value is located at No. 6. In other words, the lower the $\hat{d}$ value and the longer the edge that intersects the line, the greater the error.

In other areas, the RMSE was in the range of 1300–3600, and the MAPE values were reduced by 0.3%–1.0%. However, in the region containing the depot (No. 9), the performance is worse because the adjustment of the tour length is sufficient and the linehaul distance is over-adjusted. These results indicate that our proposed method of calculating the linehaul distance is useful for regions where the depot is far from the service area, contributing to the improvement in the estimation accuracy.

7			8			9		
	RMSE	MAPE		RMSE	MAPE		RMSE	MAPE
(4)	9534.3	2.3 %	(4)	8803.4	3.3 %	(4)	8210.7	6.1 %
(16)	5945.0	1.4 %	(16)	6063.7	2.3 %	(16)	9031.0	7.1 %

4			5			6		
	RMSE	MAPE		RMSE	MAPE		RMSE	MAPE
(4)	9584.3	2.2 %	(4)	8055.7	2.9 %	(4)	11613.6	6.5 %
(16)	5957.1	1.3 %	(16)	5559.7	2.0 %	(16)	6283.2	3.4 %

1			2			3		
	RMSE	MAPE		RMSE	MAPE		RMSE	MAPE
(4)	8704.2	1.8 %	(4)	6474.8	1.9 %	(4)	8522.0	3.4 %
(16)	5655.3	1.1 %	(16)	5010.0	1.6 %	(16)	7206.8	2.8 %

Fig. 6. Model fit comparison by layout pattern of depot and service area (Models (4) and (16))

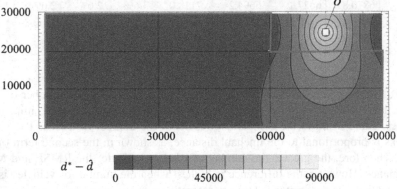

Fig. 7. Mapping of difference values between $\hat{d}$ and d^*.

5   Conclusions

In this study, we analyzed approximations of the linehaul distance between the depot and rectangular service area to improve the accuracy of estimating the total VRP distance. The VRP approximation formula is useful for the strategic and planning analyses of transportation and logistics problems, where the number and location of customers change daily. We defined a continuous approximation formula to find the average distance of a right triangle with the depot as an acute angle and the average distance between a point and a rectangular region at any given location. This approximation formula was used as the linear distance in the VRP model. In addition, we computed the optimal route for many instances with different demands, number of trucks, and locations that can be considered in the VRP approximation model. The numerical solutions were estimated using the approximation model. The results showed an improvement in the accuracy of the approximation equation for service areas far from the depot. Our results can be applied when planning deliveries, for cases where the depot is located at the edge of the city and outside the delivery area and to the planning of area segmentation represented by the strip strategy (e.g., Daganzo, Franceschetti et al. [4, 8]). The linehaul distance approximation is also useful for estimating the distance traveled by cabs and kickboards without tours. In the future, we plan to apply the developed model to analyze the above problems, while considering the time constraints of the VRP and the effects of area segmentation on the delivery efficiency.

Acknowledgments. This work was supported by JSPS KAKENHI Grant Numbers JP21K14314 and Obayashi Foundation. We appreciate their support.

References

1. Beardwood, J., Halton, J.H., Hammersley, J.M.: The shortest path through many points. Math. Proc. Camb. Philos. Soc. **55**(4), 299–327 (1959). https://doi.org/10.1017/S03050041 00034095
2. Chien, T.W.: Operational estimators for the length of a traveling salesman tour. Comput. Oper. Res. **19**(6), 469–478 (1992). https://doi.org/10.1016/0305-0548(92)90002-M
3. Cook, W.J., Applegate, D.L., Bixby, R.E., Chvátal, V.: The Traveling Salesman Problem. Princeton University Press, Princeton (2011)
4. Daganzo, C.F.: The distance traveled to visit N Points with a maximum of C Stops per vehicle: an analytic model and an application. Transp. Sci. **18**(4), 331–350 (1984). https://doi.org/10. 1287/trsc.18,4,331
5. Daganzo, C.F.: The length of tours in zones of different shapes. Transp. Res. Part B **18**(2), 135–145 (1984). https://doi.org/10.1016/0191-2615(84)90027-4
6. Eilon, S., Watson-Gandy, C.D.T., Christofides, N.: Distribution Management: Mathematical Modelling and Practical Analysis. Charles Griffin, London (1971). https://doi.org/10.1109/ TSMC.1974.4309370
7. Figliozzi, M.A.: Planning approximations to the average length of vehicle routing problems with varying customer demands and routing constraints. Transp. Res. Rec. **2089**(1), 1–8 (2008). https://doi.org/10.3141/2089-01
8. Franceschetti, A., Honhon, D., Laporte, G., Van Woensel, T.V., Fransoo, J.C.: Strategic fleet planning for city logistics. Transp. Res. Part B Methodol. **95**, 19–40 (2017). https://doi.org/ 10.1016/j.trb.2016.10.005
9. Huang, M., Smilowitz, K.R., Balcik, B.: A continuous approximation approach for assessment routing in disaster relief. Transp. Res. Part B Methodol. **50**, 20–41 (2013). https://doi.org/10. 1016/j.trb.2013.01.005
10. Jabali, O., Gendreau, M., Laporte, G.: A continuous approximation model for the fleet composition problem. Transp. Res. Part B Methodol. **46**(10), 1591–1606 (2012). https://doi.org/ 10.1016/j.trb.2012.06.004
11. Jaillet, P.: A priori Solution of a traveling salesman problem in which a random subset of the customers are visited. Oper. Res. **36**(6), 929–936 (1988). https://doi.org/10.1287/opre.36. 6.929
12. Koshizuka, T., Kurita, O.: Approximate formulas of average distances associated with regions and their applications to location problems. Math. Program. **52**(1–3), 99–123 (1991)
13. Koshizuka, T.: Integral Geometry for Applications - Measures of Figures. Kindai Kagaku Sha Co., Ltd. (2019). (in Japanese)

14. Kurita, O.: Mathematical Models of Cities and Regions - Mathematical Methods in Urban Analysis. Kyoritsu Shuppan Co., Ltd. (2013). (in Japanese)
15. Newell, G.F., Daganzo, C.F.: Design of multiple-vehicle delivery tours—I a ring-radial network. Transp. Res. Part B **20**(5), 345–363 (1986). https://doi.org/10.1016/0191-2615(86)90008-1
16. Robusté, F., Estrada, M., López-Pita, A.: Formulas for estimating average distance traveled in vehicle routing problems in elliptic zones. Transp. Res. Rec. **1873**(1), 64–69 (2004). https://doi.org/10.3141/1873-08
17. Solomon, M.M.: Algorithms for the vehicle routing and scheduling problems with time window constraints. Oper. Res. **35**(2), 254–265 (1987). https://doi.org/10.1287/opre.35.2.254
18. Stein, D.M.: An asymptotic, probabilistic analysis of a routing problem. Math. Oper. Res. **3**(2), 89–101 (1978). https://doi.org/10.1287/moor.3.2.89
19. CVRPLIB. http://vrp.atd-lab.inf.puc-rio.br/index.php/en/. Accessed 20 Jan 2022
20. TSPLIB. http://comopt.ifi.uni-heidelberg.de/software/TSPLIB95/. Accessed 20 Jan 2022
21. The instances of Solomon. http://web.cba.neu.edu/~msolomon/problems.htm. Accessed 20 Jan 2022
22. Benchmark–Capacitated vehicle routing problem (CVRP), Localsolver. https://www.localsolver.com/benchmarkcvrp.html. Accessed 20 Jan 2022

Optimal Delivery Area Assignment for the Capital Vehicle Routing Problem Based on a Maximum Likelihood Approach

Junya Maruyama[1] , Yudai Honma[1]([✉]) , Daisuke Hasegawa[1] , Soma Toki[2] ,
and Naoshi Shiono[3]

[1] The University of Tokyo, Bunkyo, Tokyo, Japan
`juntama0826@g.ecc.u-tokyo.ac.jp`, `{yudai,`
`hasega60}@iis.u-tokyo.ac.jp`
[2] Tokyo Gas Co., Ltd., Minato, Tokyo, Japan
`toki.s@tokyo-gas.co.jp`
[3] Kanagawa Institute of Technology, Atsugi, Kanagawa, Japan
`na-shiono@ic.kanagawa-it.ac.jp`

Abstract. In this study, we constructed an optimization model for the maximum likelihood estimation of delivery areas from a capacitated vehicle routing problem. The aim is to develop a method that combines the advantages of two methods of delivery planning: the efficiency of the routing software-based method and the flexibility of the area-in-charge method. We first conduct computer experiments to derive the optimal cycling plan for each stochastic demand pattern. We then solve the optimal delivery area assignment that is globally consistent with the data from these experiments. We focused on whether the optimal route for each demand pattern was contained in the same area and found the assigning area that maximized the probability. This model is designed for daily use because it is an easy-to-interpret area map, while the optimization of the circulation problem is solved using computers in advance. In experiments using the data, we confirmed that the model can provide correct area creation.

Keywords: Delivery plan · Area assignment · Maximum likelihood estimation

1 Introduction

Planning an effective delivery route is one of the most important topics discussed in logistics, such as postal delivery services and meter readings. In recent years, its importance has increased. Figure 1 shows the distribution of customers for gas cylinder delivery. About 5,000 customers exist. Distributors deliver 30 to 70 cylinders every day. That is, daily demand points are stochastic. For distributors, there are mainly two effective strategies for stochastic delivery route planning: one is to find the best route through routing software, and the other is to assign the driver's area to each driver.

The routing software-based method of determining the optimal route is more efficient than the method of assigning a driver's area. The method of searching for the optimal

© The Author(s), under exclusive license to Springer Nature Switzerland AG 2022
B. Dorronsoro et al. (Eds.): OLA 2022, CCIS 1684, pp. 157–167, 2022.
https://doi.org/10.1007/978-3-031-22039-5_13

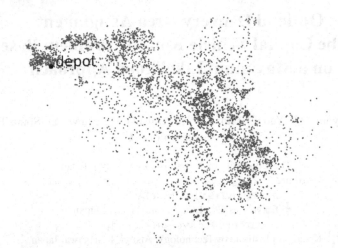

Fig. 1. Demand points

route has been studied as the Vehicle Routing Problem (VRP). On the other hand, solving the VRP is computationally demanding, so it is not practical to find the optimal route. In actual delivery, distributors face sudden customer changes, such as a customer's request due to running out of gas or bad weather. In these cases, a manager change driver's scheduling by hang as applying the software takes much time.

Alternatively, in daily delivery, plans are often made based on the area assignment for each driver. For many distributors, these areas are determined by discussions among drivers or by the experience of managers without any actual quantitative analysis. Assignment areas for each driver are often inefficient. On the other hand, such a delivery plan has the advantage of being able to respond flexibly to sudden changes in the delivery destination or the absence thereof. Such irregular events are common in everyday delivery.

As shown above, the two methods have trade-offs in terms of computational time, traveling time, and flexibility in actual daily delivery when we make a delivery route plan.

Previous attempts in logistics have been made to assign areas appropriately. In the planning and strategic phases, a successive approximation approach with simple assumptions (uniform random demand distribution, simple shapes) estimated the impact of zoning. Newell and Daganzo [1] analyzed the effect on distance when a rectangular area was divided into equal parts and found the optimal number of divisions. Applying these approaches, Ouyang [2] proposed an algorithm in which geometric features determine the vehicle routing zone for any demand distribution. In a more practical approach, Galvão et al. [3] focused on the density distribution of demand and tried to determine the area using the Voronoi diagram approach. They relaxed the predetermined boundaries of the partition and iteratively modified them until they converged to minimize the distance traveled by vehicles. Ayala et al. [4] developed a demand point allocation scheme that minimizes the distance from the depot to the network connecting the demand point to the depot or the demand points to each other. Zhong et al. [5] divided the field into areas by preparing cells, which are sets of demand points, and considered the optimal clustering of cells by adding a cost to each. In recent research, Sung et al. [6] used a meta-heuristics approach to zoning aerial vehicles by bearing from depots to demand points. However, there is no study that solves the VRP for the entire field and naturally divides the distributor's area into driver's areas.

The purpose of this research is to propose a new optimization model to find the area assignment for the capacitated vehicle routing problem, which maintains both efficiency (by calculating the optimal route) and flexibility (by defining the area assignment). To achieve this purpose, we first conducted computer experiments to derive the optimal cycling plan for each stochastic demand pattern. We then identified the optimal delivery area assignment that was globally consistent with the data from these experiments. We introduce a clustering method in the work of Honma et al. [7], which is a maximum likelihood estimation of areas from transition information.

One of the features of this research is that we aimed to produce output in the form of a map for use in actual delivery scenarios. It is frequently difficult to handle and calculate complex mathematical models in the field. Even in such cases, the output in the form of a map provides an easy and intuitive way to use the results of this research. In addition, because of the map format, the delivery personnel do not have to significantly alter what they have been doing. Thus, it can be said that the output of this research can achieve efficiency with less effort.

In addition, by using area assignment, the system can handle demands that occur randomly every day. In actual delivery, demand occurs randomly among a number of potential demand points, and it is necessary to create a delivery plan that corresponds to the demand. This is much easier to operate than finding a solution through software.

This paper is organized as follows. In Sect. 2, we formulate the problem. We propose a mathematical model for achieving efficient area assignment, which is the objective of this study and is estimated by taking the maximum likelihood from the optimal route and constructing a mathematical model for assigning an area. In Sect. 3, we introduce the data for the computer experiments used to test the certainty of the model, and in Sect. 4, we verify the model use actual data. Section 5 discusses the model based on the results of Sect. 4, and finally, Sect. 6 provides conclusions and future perspectives for further study.

2 Formulation

2.1 Our Concept

In this study, we deal with the capacitated Vehicle Routing Problem (CVRP). To assign the optimal area for each driver, we iteratively calculate the optimal traveling routes on software under stochastic demands and then estimate the assignment areas based on the solutions to be as consistent as possible with the computer experiments. This means that we assign a label of area information for each demand point. Thereby, this problem can be rephrased as the problem of labeling area information at the demand points on the same circular route in computer experiments in as few areas as possible.

First, we present the conditions for achieving this objective mathematically. In this study, we consider an area assignment that satisfies the following conditions that an area is as consistent as possible.

i. Demand points that belong to the same area should be delivered by the same vehicle as much as possible.
ii. Demand points that belong to different areas should be delivered by different vehicles as much as possible.

These show the properties that an optimal area assignment should satisfy. For these properties, we conduct a large number of computer experiments because we assume that daily demand points are random. Therefore, we apply probabilities i and ii. An area assignment that satisfies these properties is achieved by combining the following procedures, each corresponding to the above conditions:

i'. Maximize the probability of delivering any given demand points in the same area with the same vehicle
ii'. Minimize the probability of delivering any given demand points in different areas with the same vehicle

By achieving these conditions at the same time, the solution should be obtained such that the same drivers deliver as much as possible within the same area, conversely, the same drivers do not deliver as much as possible in different areas. When solving our area assignment problem, a constraint is employed in that each vehicle has the same capacity. This is to keep the quantity transported by each delivery vehicle constant. The optimization problem is formulated as a maximum likelihood estimation by partitioning a field as described above.

2.2 Formulation Based on Demand Points

Our area assignment problem is solved as a quadratic assignment problem in terms of the label of area information [8].

Let I be the set of demand points and A be the set of areas for all drivers. The presence or absence of delivery to demand points $i(\in I)$ and $j(\in I)$ is determined by repeated computer experiments. We now introduce a 0–1 variable as follows:

$$z_{ia} = \begin{cases} 1 \text{ (If demand point i belongs to area } a(\in A)) \\ 0 \text{ (otherwise)} \end{cases} \tag{1}$$

where the number of areas is determined by the number of vehicles used for daily delivery. The purpose of this study is to divide the entire target field into areas without overlapping by multiple areas and to assign a demand point to one area. Then, the following relationship is established:

$$\sum_{a \in A} z_{ia} = 1 \quad \forall i \in I \tag{2}$$

Let K be the set of trials for computer experiments and I_k be the set of demand points at which demand points occur in a certain trial $k(\in K)$. We denote by $C(I_k)$ the set of functions of I_k that extracts from I_k every combination that takes two demand points. Then, $z_{ik}z_{jk}$ indicates the presence or absence of a path connecting any demand points i and j belonging to an area $a(\in A)$ in a certain trial $k(\in K)$. The total number of paths between demand points belonging to the same area in all trials is expressed as follows:

$$\sum_{a \in A} \sum_{(i,j) \in C(I_k)} \sum_{k \in K} z_{ia}z_{ja} \tag{3}$$

The path across the different areas, except for the upper one, can be expressed as follows:

$$\sum_{a \in A} \sum_{(i,j) \in C(I_k)} \sum_{k \in K} (1 - z_{ia}z_{ja}) \tag{4}$$

In addition, we also introduce the following 0–1 variables:

$$s_{ij}^k = \begin{cases} 1 & \begin{array}{l} \text{(If demand points i and j are delivered} \\ \text{by the same vehicle in trial } k(\in K)) \end{array} \\ 0 & \text{(otherwise)} \end{cases} \tag{5}$$

Using these variables, the total number of routes delivered to the same vehicle in each area can be shown as follows:

$$\sum_{a \in A} \sum_{(i,j) \in C(I_k)} \sum_{k \in K} s_{ij}^k \times z_{ia}z_{ja} \tag{6}$$

Similarly, the total number of deliveries between any demand points in different areas by the same vehicle can be shown as follows:

$$\sum_{a \in A} \sum_{(i,j) \in C(I_k)} \sum_{k \in K} s_{ij}^k \times (1 - z_{ia}z_{ja}) \tag{7}$$

The probability f_{in} of delivering to any demand point in the same area with the same vehicle and the probability f_{out} of delivering to any demand point in a different area with the same vehicle, which are the conditions that appeared in Sect. 2.1, can be expressed using the above variables as follows:

$$f_{in} = \frac{\sum_{a \in A} \sum_{(i,j) \in C(I_k)} \sum_{k \in K} s_{ij}^k \times z_{ia}z_{ja}}{\sum_{a \in A} \sum_{(i,j) \in C(I_k)} \sum_{k \in K} z_{ia}z_{ja}} \tag{8}$$

$$f_{out} = \frac{\sum_{a \in A} \sum_{(i,j) \in C(I_k)} \sum_{k \in K} s_{ij}^k \times (1 - z_{ia}z_{ja})}{\sum_{a \in A} \sum_{(i,j) \in C(I_k)} \sum_{k \in K} (1 - z_{ia}z_{ja})} \tag{9}$$

Furthermore, in this mathematical optimization, we set the constraint for capacity, i.e., the number of demand points belonging to each area, using the number of vehicles N as follows. The purpose of this is to adjust the delivery volume of each vehicle for practical applications:

$$\left\lfloor \frac{|I|}{N} \right\rfloor \leq \sum_{i \in I} z_{ia} \leq \left\lceil \frac{|I|}{N} \right\rceil \qquad \forall a \in A \tag{10}$$

Since satisfying conditions i and ii in Sect. 2.1 simultaneously means minimizing $f_{in} - f_{out}$, the following mathematical optimization problem is established:

$$max \quad f_{in} - f_{out} \tag{11}$$

$$s.t. \quad \sum_{a \in A} z_{ia} = 1 \quad \forall i \in I \tag{12}$$

$$\left\lfloor \frac{|I|}{N} \right\rfloor \le \sum_{i \in I} z_{ia} \le \left\lceil \frac{|I|}{N} \right\rceil \quad \forall a \in A \tag{13}$$

$$z_{ia}, s_{ij}^k \in \{0, 1\} \quad \forall i, j \in I, \forall a \in A, \forall k \in K \tag{14}$$

2.3 Formulation Based on Zones

In Sect. 2.2, we formulated the equation based on the demand points. However, there are two issues. First, quadratic assignment problems with over 5,000 points cannot be solved in a short time. Second, considering practical applications, it is more convenient for the same vehicles to visit the same town.

For these reasons, we introduce "zone" aggregated demand points. The concept of the zone is illustrated in Fig. 2. The zones are sets of demand points. The number of demand points for each zone is nearly equivalent. For example, zones are created by zip code. In addition, zones can be obtained by solving the p-median problem [9]. In this study, 100 zones are created, and each demand point is assigned to one zone. In the previous Sect. 2.2, we formulated the equation in such a way that demand points are allocated to areas, and in this Sect. 2.3, we formulate it in such a way that zones are allocated to areas.

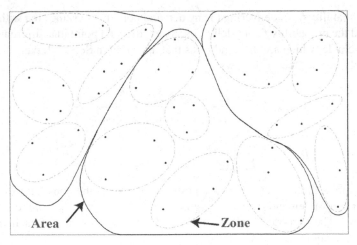

Fig. 2. Concept of zone

Let P be the set of zones. To propose a formulation that includes zones, we first introduce $p(\in P)$, and then we introduce a 0–1 variable that indicates whether zone p belongs to area $a(\in A)$:

$$x_{pa} = \begin{cases} 1 \text{ (If demand zone p belongs to area } a(\in A)) \\ 0 \text{ (otherwise)} \end{cases} \tag{15}$$

As shown in Sect. 2.2, we denote by $C(P)$ the set of functions of P that extracts from P every combination that takes two demand points. At this time, let n_{pq}^k be the number of combinations such that one demand point is in zone p and another is in zone q among all combinations $C(I_k)$ of demand points I_k visited in trial k. The total number of deliveries in zones p and q of the same area a patrolled in a certain trial k can be expressed as $n_{pq}^k \times x_{pa} x_{qa}$, and the total number of deliveries within that same area is:

$$\sum_{a \in A} \sum_{(p,q) \in C(P)} \sum_{k \in K} n_{pq}^k \times x_{pa} x_{qa} \tag{16}$$

As in Sect. 2.2, the total number of routes that cross different areas is as follows:

$$\sum_{a \in A} \sum_{(p,q) \in C(P)} \sum_{k \in K} n_{pq}^k \times \left(1 - x_{pa} x_{qa}\right) \tag{17}$$

Furthermore, let t_{pq}^k be the number of combinations of all combinations $C(I_k)$ of demand points I_k visited in trial k, where one demand point is in zone p and the other in zone q, and where they are serviced by the same vehicle. Among those whose routes are in the same area, the total number delivered by the same vehicle is expressed:

$$\sum_{a \in A} \sum_{(p,q) \in C(P)} \sum_{k \in K} t_{pq}^k \times x_{pa} x_{qa} \tag{18}$$

In the same way as in Sect. 2.2, those items whose route is not in the same area, and which are delivered by the same vehicle, are shown as follows:

$$\sum_{a \in A} \sum_{(p,q) \in C(P)} \sum_{k \in K} t_{pq}^k \times \left(1 - x_{pa} x_{qa}\right) \tag{19}$$

Using the above, the f_{in} of delivering to any zone in the same area with a same vehicle and the probability and f_{out} of delivering to any zone in a different area with the same vehicle, which replace demand points to zones as they appeared in Sect. 2.2, can be shown as follows:

$$f_{in} = \frac{\sum_{a \in A} \sum_{(p,q) \in C(P)} \sum_{k \in K} t_{pq}^k \times x_{pa} x_{qa}}{\sum_{a \in A} \sum_{(p,q) \in C(P)} \sum_{k \in K} n_{pq}^k \times x_{pa} x_{qa}} \tag{20}$$

$$f_{out} = \frac{\sum_{a \in A} \sum_{(p,q) \in C(P)} \sum_{k \in K} t_{pq}^k \times \left(1 - x_{pa} x_{qa}\right)}{\sum_{a \in A} \sum_{(p,q) \in C(P)} \sum_{k \in K} n_{pq}^k \times \left(1 - x_{pa} x_{qa}\right)} \tag{21}$$

When the number of zones is constant, the number of demand points will also be constant, thus guaranteeing that the capacity of each zone is constant. Using this fact, and adding that the number of each zone should be as equal as possible, so that $f_{in} - f_{out}$ is maximized, the mathematical optimization problem can be shown as follows:

$$max \quad f_{in} - f_{out} \tag{22}$$

$$s.t. \quad \sum_{a \in A} x_{pa} = 1 \quad \forall p \in P \tag{23}$$

$$\left\lfloor \frac{|I|}{N} \right\rfloor \leq \sum_{p \in P} x_{pa} \leq \left\lceil \frac{|P|}{N} \right\rceil \quad \forall a \in A \tag{24}$$

$$x_{pa} \in \{0, 1\} \forall p \in P, \forall a \in A \tag{25}$$

3 Data Preparation

The optimization problem formulated in Sect. 2.3 was solved using the following experimental data. We considered the distribution of demand points as shown in Fig. 1. There are 5,000 candidate demand points. There is only one depot in the upper left, and all delivery vehicles originate from here.

First, we randomly select demand points in accordance with the number of vehicles and the number of demand points delivered by one vehicle in one day. This means the distribution of demand for a hypothetical day. Next, for these demand points, we solve the CVRP using the LocalSolver package [10]. This is equivalent to calculating the optimal route minimizing the total distance for one day's demand distribution. We repeat these procedures for 300 trials, which means one year of delivery operations. In this way, computational experiments were conducted to find the optimal route for a year of demand. Thus, we obtained 300 patterns of hypothetical optimal delivery plans as computer experiments for solving the optimal problem.

By overlaying these computer experiment data, we can intuitively grasp the area distribution as shown in Fig. 3. This figure shows a case with six delivery vehicles, and routes delivered by the same vehicle are shown in the same color. Such areas are created by the maximum likelihood estimated in (22)–(25).

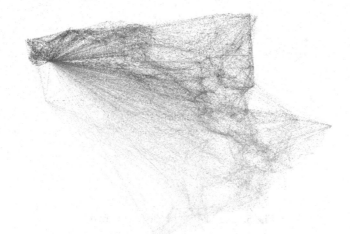

Fig. 3. Assembly of routes

4 Results

The following are the results of obtaining the solution using LocalSolver 10.5 [11] with the equation in Sect. 2.3 and the data prepared in Sect. 3. The area assignment of demand points for the cases of 6 and 14 vehicles is shown in Fig. 4 and Fig. 5 so that the same area has the same color. In addition, the white dot in the figures is the depot (starting point) of the delivery.

Fig. 4. Area solve (6 cars = 6 areas)

Fig. 5. Area solve (14 cars = 14 areas)

5   Discussion

From the results in the previous section, we can see that the formulation of this study allows us to make a clear assignment of areas, thus confirming that the area assignment was reasonable. These results can be obtained regardless of the number of vehicles. Therefore, we can say that the formulation and constraints used in the process are appropriate. On the other hand, since we imposed a very strict condition on the capacity, it is necessary to examine how it changes when this condition is relaxed.

In this area assignment, the area spreads radially. This is an obvious solution since all the computer experiments start from the depot, but it also works positively in terms of efficiency in the actual delivery scenarios. In this sense, we can say that we have succeeded in creating a map that is suitable for the actual delivery scenarios.

6   Conclusion

In this study, we developed a model for the maximum likelihood estimation of area assignment from optimal routes. From the results, it can be inferred that the model is appropriate. In the next step, it will be necessary to verify whether the area assignment obtained in this study is superior in terms of efficiency to the currently used delivery areas.

References

1. Newell, F., Gordon, D., Carlos, F.: Design of multiple vehicle delivery tours-II other metrics. Transp. Res B-Meth **20**(5), 365–376 (1986). https://doi.org/10.1016/0191-2615(86)90009-3
2. Ouyang, Y.: Design of vehicle routing zones for large-scale distribution systems. Transp. Res B-Meth: Methodol. **41**(10), 1079–1093 (2007). https://doi.org/10.1016/j.trb.2007.04.010
3. Galvão, L.C., Novaes, A.G.N., Souza de Cursi, J.E., Souza, J.C.: A multiplicatively-weighted Voronoi diagram approach to logistics districting. Comput. Oper. Res. **33**, 93–114 (2006). https://doi.org/10.1016/j.cor.2004.07.001

4. García-Ayala, G., González-Velarde, J.L., Ríos-Mercado, R.Z., Fernández, E.: A novel model for arc territory design: promoting Eulerian districts. Int. Trans. Oper. Res. **23**, 433–458 (2016). https://doi.org/10.1111/itor.12219

5. Zhong, H., Hall, R.W., Dessouky, M.: Territory planning and vehicle dispatching with driver learning. Transp. Sci. **4**(1), 74–89 (2007). https://doi.org/10.1287/trsc.1060.0167

6. Sung, I., Nielsen, P.: Zoning a service area of unmanned aerial vehicles for package delivery services. J. Intell. Rob. Syst. **97**(3–4), 719–731 (2019). https://doi.org/10.1007/s10846-019-01045-7

7. Yudai, H., Atsushi, S.: New clustering method to estimate overlapped exchange area from regional flow matrix –characteristics of population migrations and freight flow. J. City Plan. Inst. Japan **55**(3), 475–481 (2020). https://doi.org/10.11361/journalcpij.55.475

8. Mikio, K., Pedroso, J.P., Masakazu, M., Rais, A.: Mathematical Optimization: Solving Problems Using Gurobi and Python, 2nd edn. Kindai Kagaku sha Co., Ltd., Japan (2013)

9. Densham, P.J., Rushton, G.: A more efficient heuristic for solving large p-median problems. J. RSAI **71**(3), 307–329 (1992). https://doi.org/10.1007/BF01434270

10. Capacitated Vehicle Routing (CVRP) package, LocalSolver. https://www.localsolver.com/docs/last/exampletour/vrp.html. Accessed 20 Jan 2022

11. LocalSolver. https://www.localsolver.com/. Accessed 20 Jan 2022

Neural Order-First Split-Second Algorithm for the Capacitated Vehicle Routing Problem

Ali Yaddaden[✉], Sébastien Harispe, and Michel Vasquez

EuroMov Digital Health in Motion, Univ Montpellier, IMT Mines Ales, Ales, France
{ali.yaddaden,sebastien.harispe,michel.vasquez}@mines-ales.fr

Abstract. Modern machine learning, including deep learning models and reinforcement learning techniques, have proven effective for solving difficult combinatorial optimization problems without relying on handcrafted heuristics. In this work, we present NOFSS, a Neural Order-First Split-Second deep reinforcement learning approach for the Capacity Constrained Vehicle Routing Problem (CVRP). NOFSS consists of a hybridization between a deep neural network model and a dynamic programming shortest path algorithm (Split). Our results, based on intensive experiments with several neural network model architectures, show that such a two-step hybridization enables learning of implicit algorithms (i.e. policies) producing competitive solutions for the CVRP.

Keywords: Neural combinatorial optimization · Capacitated vehicle routing problem · Order-first split-second · Deep reinforcement learning

1 Introduction

Modern machine learning, including deep learning models and reinforcement learning techniques, have proven effective for solving difficult combinatorial optimization problems without relying on handcrafted heuristics [1]. The framework known as Neural Combinatorial Optimization (NCO), which proposes to solve combinatorial optimization problems using recent neural networks architectures, is in this context widely studied for routing problems such as the traveling salesman problem (TSP) [2–5] and the capacitated vehicle routing problem (CVRP) [5,6].

Current NCO approaches implement a construction-based strategy. For the CVRP, such approaches build (i.e. construct) candidate solutions step by step, by selecting at each time step either to visit a client or to go back to the depot to refill, until each client is served. The action to perform at each construction step is chosen based on a probability distribution that will be estimated by a deep neural network, either using supervised or reinforcement learning. This

---

S. Harispe—This work used HPC resources of IDRIS (allocation 2022-AD011011309R2) made by GENCI.

discrete probability distribution defines the probability that an extension of the partial solution under construction, considering each available choices (unsatisfied clients and depot), will lead to the optimal solution. Considering such construction-based NCO approaches, solving the CVRP is therefore reframed as a learning goal aiming to obtain a good estimate of the probability distribution, such as step decisions based on this estimate minimize solution costs.

Using such an approach, the models handle both clients routing and returns to depot. In this context, choices of when to return to the depot are critical. Indeed, more returns to the depot can *de facto* lead to candidate solutions with a number of tours[1] greater than the optimal one. This will result in models failing to efficiently learn interesting resolution strategies, i.e. routing *policies*, due to poor quality candidate solutions, and/or large computational costs inducing prohibitive learning process (millions of learning steps). Handcrafted heuristics and metaheuristics may nevertheless be used to handle return to depot by using an exact tour splitting algorithm - solving a shortest path problem in an auxiliary graph that represents the clients' visit order [7,8]. Inspired by this problem decomposition, this paper presents NOFSS, Neural Order-First Split-Second, a novel two-step learning-based approach proposing to:

1. Learn how to order clients into a giant tour, using a deep neural network.
2. Optimally split the giant tour into a feasible solution using an exact split algorithm.

NOFSS is a generic approach that will be introduced and tested in the context of CVRP, even if it may be used for a larger class of routing problems. NOFSS relies on a deep neural network that learns a giant tour policy and a dynamic programming algorithm, called Split [8]. Split modifies the giant tour into a feasible solution with respect to vehicle capacity and clients demands. It acts as an oracle that provides feedback on the quality (the total travelled distance) of the giant tour generated from our neural network. This makes it possible to train the NOFSS model through REINFORCE algorithm.

Alongside NOFSS introduction, we present an extensive comparison of various NOFSS and NCO models with state-of-the-art CVRP (meta)heuristics. Results show that, by exploring the search space of giant tours, NOFSS allows to implicitly learn competitive routing policies.[2]

The paper is organized as follows: Sect. 2 formally introduces the CVRP and notations; Sect. 3 introduces related work focusing on approaches based on machine learning; Sect. 4 presents NOFSS; Sect. 5 presents the experimental protocol as well as results. Discussions and perspectives conclude the paper.

---

[1] A tour is the ordering of clients the vehicle will visit before returning back to the depot. The optimal number of tours will therefore depend on client's demands and vehicle capacity.

[2] Our implementation and results will be available on the following repository https:// github.com/AYaddaden/NOFSS.

2 Problem Statement

The Capacitated Vehicle Routing Problem (CVRP) is one of the basic types of routing problems where information associated with the clients, the depot and the vehicles are deterministic and known in advance. We consider a set of n clients dispatched on the Euclidean plan and a single depot. In the depot, there is a fleet of homogeneous vehicles with identical transport capacity C. We associate to the clients their coordinates (x_i, y_i) and their demands of goods to deliver $0 \leq d_i \leq C$ $(i \in \{1, ..., n\})$. We associate to the depot its coordinates (x_0, y_0). The demands cannot be split, meaning that a vehicle must satisfy the demand at once. The objective is to minimize the total travelled distance when serving all the clients.

The problem can also be formulated using graph theory [9]. We consider a complete graph $G(V, E)$, where $V = \{0, ..., n\}$ is the vertex set (the vertex 0 represents the depot) and $E = \{(u, v) \in V \times V, u \neq v\}$ is the edge set. We associate with each edge a cost defined as the distance between two vertices. We can represent it as a cost matrix D where $D_{uv} = \sqrt{(x_u - x_v)^2 + (y_u - y_v)^2}$, $(u, v) \in E$. The goal is in this case to find simple circuits called tours such that all clients are served without transgressing the vehicles' capacity and the total travelled distance is as minimum as possible.

3 Related Work

3.1 Neural Combinatorial Optimization (NCO) for the CVRP

We refer to the use of end-to-end deep neural network approaches for solving difficult combinatorial optimization as the Neural Combinatorial Optimization (NCO) framework [3]. In this section, we review the use of this framework to learn construction-based policies for routing problems.

Although the use of neural networks for solving combinatorial optimization problems dates back longer than the appearance of modern deep learning architectures [10], their use has faded away in favor of more efficient metaheuristics. The success of deep learning and reinforcement learning has revived the interest in studying deep neural networks for solving this class of problems. More precisely, with the appearance of the sequence-to-sequence type approaches and the attention mechanism. The general framework (Fig. 1) considers two neural networks called respectively encoder and decoder, which can be of different types. The encoder generates the *embeddings* of each element of a problem instance (clients and depot). Embeddings can be viewed as an alternative representation of the element in a higher dimension vector space ($\mathbb{R}^d$ with generally $d = 64$ or $d = 128$). This representation is intended to encompass meaningful features that will be used during the decoding phase. The decoder uses the history of the already visited elements (clients or depot) to compute a query vector that summarizes the solution under construction through a single vector. The query along with the embeddings are used to compute a probability distribution of selecting the next element via an attention module. To do so, the attention module confronts the query $q \in \mathbb{R}^d$ to the elements embeddings $e_i \in \mathbb{R}^d$ in order to give attention scores s_i either via a scaled dot-product (i.e. $s_i = \frac{q \cdot e_i^T}{\sqrt{d}}$) or via an additive attention defined as $s_i = v^T \cdot tanh(W_q \cdot q + W_e \cdot e_i)$ with $W_q, W_e \in \mathbb{R}^{d \times d}$, $v \in \mathbb{R}^d$ being learnable parameters. The scores s_i will be converted into a probability distribution by a softmax function[3].

Pointer Networks [2] was the seminal work that considered training LSTM-based encoder and decoder along with an additive attention module via supervised learning on a dataset of TSP instances. The approach successfully solved instances of sizes between 10 and 50 cities. It was next improved by using a policy-based reinforcement learning algorithm for training, namely REIN-FORCE with critic baseline, thus avoiding the need of a supervision, i.e. to have ground truth optimal solutions for the TSP dataset's instances [3]. Reinforcement learning proved to be more effective for training models on instances of size between 20 and 100 cities, thus achieving better results than Pointer Networks.

Nazari et al. [6] applied the NCO approach to CVRP. Their model considered 1D convolutions instead of an LSTM encoder in order not to bias the model on the inputs' order – LSTM are indeed better suited for modeling sequences where input's order matters. Comparison with classic CVRP algorithms (Clarke and Wright *savings* heuristic and the Sweep algorithm) shows that the deep neural network model performs better on training and test instances' sizes ranging from 10 to 100 clients. It appears also, that the choices of the encoder and decoder are of extreme importance in order to improve the learned policy. The Attention Model (AM) improves the results on the TSP and the CVRP by introducing a model entirely based on the attention mechanism [5]. It uses a Transformer encoder and computes the query vector using a Multi-head attention [11]. The Transformer encoder allows taking into account the graph structure of the TSP and the CVRP in the same way Graph Neural Networks do, thus giving a better representation of the instances. Also, they introduce a new baseline for the REINFORCE algorithm; a greedy rollout baseline that is a copy of AM that gets updated less often.

3.2   Two-Step Algorithms for the Vehicle Routing Problem

Classical two-step construction approaches for solving the CVRP involve (i) partitioning the clients into feasible clusters with regard to vehicle capacity and (ii) ordering them into routes of minimum length. Based on how the two operations are orchestrated, we can distinguish two types of two-step algorithms: Cluster-first Route-second and Order-first Split-Second.

In Cluster-first Route-second algorithms, the clients are first grouped together following the vehicle capacity constraint, then a traveling salesman problem is solved for each cluster using an exact solver or heuristics. The Sweep algorithm is the most common algorithm of this type [12]. Feasible clusters are

---

[3] $softmax(s_i) = \frac{exp(s_i)}{\sum_{j=1}^{K} exp(s_j)}$.

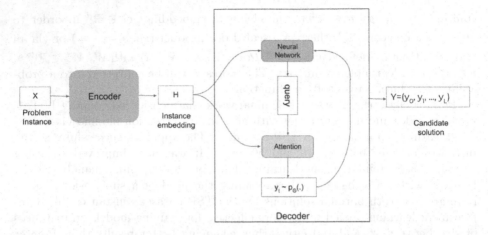

Fig. 1. The general encoder-decoder framework used to solve routing problems. The encoder takes as input a problem instance X and outputs an alternative representation H in an embedding space. The decoder iteratively constructs the candidate solution Y by adding a client or a depot y_t at each step t until all clients are visited.

constructed by considering the polar angle between the clients and the depot, then for each cluster a TSP is solved. An extension of this algorithm called the petal algorithm considers generating several routes and selects the final routes of the solution by solving a set partitioning problem [13]. Another work considers obtaining the clusters by solving a generalized assignment problem [14]. One major drawback of this approach is that it is not computationally efficient due to the clustering algorithms [15].

On the other hand, Order-first Split-second algorithms consider first ordering the customers into a sequence called a giant tour, to then, decompose it into a set of feasible tours considering the vehicle capacity. Traveling salesman problem heuristics are used to get giant tours, and the CVRP tours can be obtained optimally from the giant tours by solving a shortest path problem, as we will detail later. The first documented approach of this type generates the giant tour by random permutation of clients' visit order, followed by a 2-opt improvement, and then builds the routes using Floyd's algorithm [7]. Prins proposed the first genetic algorithm for the CVRP that relies on the Order-first Split-second approach, which was competitive with the best metaheuristic at that time (Tabu Search) [8]. In their approach, the authors proposed a representation of the chromosomes as giant tours and introduced the *Split* procedure based on an auxiliary acyclic graph generated on top of a giant tour. Bellman's algorithm is used in order to extract the feasible routes. HGS, today's state of the art metaheuristic for the CVRP, also uses a giant tour representation and the Split algorithm [16].

The Order-first Split-second approach is appealing. A recent review of this approach surveys more than 70 research papers that build heuristics and metaheuristics to successfully solve vehicle routing problems [17]. Computationally, it is less expensive to build a giant tour and then to split it than building clusters of clients. Also, the search space is reduced to the space of giant tours instead of the direct solution representations with depot placement. As highlighted in the survey, this search space reduction does not make the optimal solution unattainable, since there is an *optimal* giant tour which corresponds to the optimal solution. In addition, for a given giant tour, only its optimal split is retained. This ensures to prevent too many poor quality solutions from appearing often.

3.3   Graph Neural Networks

Since CVRP instances can be modelled as a graph, it is interesting to use neural networks that takes advantage of this structure. This makes Graph Neural Networks (GNNs) an ideal choice to compute a representation of an instance that captures useful information for the resolution process. We define a GNN by stacking K GNN blocks. Each block k relies on message passing in order to compute the node embeddings h_u^k, $\forall u \in V$. This mechanism can be viewed as a differentiable function that computes node embeddings as follows: $h_u^k = F(h_u^{k-1}, \{h_v^{k-1}\}_{v \in \mathcal{N}(u)}, \{e(u,v)\}_{v \in \mathcal{N}(u)})$, with $\mathcal{N}(u)$ being the set of the neighbor nodes of a node $u \in V$ and $\{e(u,v)\}_{v \in \mathcal{N}(u)}$ the set of edges that link the node u to its neighbors $v \in \mathcal{N}(u)$. We use the instance features as an initial input of the first GNN block. The function F itself relies on two mechanisms: neighborhood message aggregation and node embedding update, defined as:

$$m_u^{(k)} = \text{AGGREGATE}\left(\{h_v^{(k-1)}\}_{v \in \mathcal{N}(u)}, \{e(u,v)\}_{v \in \mathcal{N}(u)}\right)$$
$$h_u^{(k)} = \text{UPDATE}(h_u^{(k-1)}, m_u^{(k)})$$

Aggregation can either be the mean, the maximum or the sum of neighbors' node embeddings. It can also be a weighted sum with weights computed using an attention mechanism [18]. It can take into consideration the edge weights of the neighboring nodes $e(u,v)$. The update function is a deep neural network that computes a new node embedding by using the message from the aggregation and the node embedding from the preceding block. Graph neural network models differ depending on the choice of the AGGREGATE and the UPDATE functions.

We can distinguish two families of GNNs: spectral and spatial GNNs. Spectral GNNs rely on spectral graph representations based on graph signal processing theory, such as GCN [19]. Spatial GNNs, such as GAT [18], exploit the graph topology. Refer to Zhou et al. for a GNN review [20].

In the next section, we describe how we use the *Split* algorithm along with the NCO framework to train GNN models for solving the CVRP.

4   The Neural Order-First Split-Second Algorithm

As mentioned in the previous section, actual NCO construction-based policies for the CVRP produce a sequence by routing the clients and choosing when to return to the depot iteratively until all clients are served. These policies may lead to more returns to depot than necessary and produce poor quality solutions. For example, a policy can decide to refill in the depot after serving each client even if the vehicle capacity allows for serving more than one client at once. Learning from poor quality solutions can slow down and hamper the learning process and produce suboptimal policies. Instead of this, we propose to let the deep neural network build an indirect solution representation via the construction of the giant tour and to delay the routes construction to the Split algorithm. Thus, our neural network implicitly learns to solve vehicle routing problem instances by exploring the space of giant tours. Alternatively, we can view the neural network's output as a permutation of the clients' visit order, which is close to what is done in works for the TSP [3,4]. This also simplifies the masking procedure used to avoid the appearance of a client twice in the solution. Another advantage of this approach is that the neural network can learn different policies depending on the variant of the vehicle routing problem (e.g. Capacitated VRP, VRP with Time Windows) without additional adaptation. The Split algorithm will handle the additional constraints, and the neural network learns the policy accordingly. Unlink other learning-based construction approaches that build a solution in a variable number of steps due to the return to the depot to refill, our neural network builds the giant tour in a fixed number of steps equal to the number of clients in the instance. Algorithm 1 presents the general approach that will be detailed afterwards.

For a given instance X of the CVRP, our neural network defines a stochastic policy that outputs the probability of generating a giant tour as a sequence Y. Using the probability chain rule, and with θ the parameters of the neural network, this policy is defined as follows:

$$P_\theta(Y|X) = \prod_{t=0}^{n-1} p_\theta(y_t|y_0, ..., y_{t-1}, X)$$

After sampling a sequence Y from P_θ, Y is then transformed into feasible routes using the Split algorithm with regard to the vehicle's capacity constraint. The Split algorithm can be viewed as an oracle that evaluates the goodness of a giant tour by returning the associated solution's total travelled distance. This evaluation makes it possible to train our deep neural network via reinforcement learning. We define the loss as the expected tour lengths of the Y sequences evaluated by the Split algorithm, i.e. $\mathcal{L}(\theta) = \mathbb{E}_{X \sim \mathcal{D}, Y \sim P_\theta(.|X)}[\mathrm{Split}(Y, X)]$. The objective is to find the best parameters θ that will output good quality sequences Y that would result on short tour lengths. For this, we rely on ADAMW as a gradient descent optimizer during training. In order to compute the gradient of the loss, we use REINFORCE with Rollout baseline [5]:

$$\nabla_\theta \mathcal{L}(\theta) = \mathbb{E}_{X \sim \mathcal{D}, Y \sim P_\theta(.|X)}\left[\left(\mathrm{Split}(Y, X) - b(X)\right)\nabla_\theta \log P_\theta(Y|X)\right]$$

The gradient $\nabla_\theta \mathcal{L}(\theta)$ is approximated using Monte Carlo sampling over a batch of B i.i.d CVRP instances as follows:

$$\nabla_\theta \mathcal{L}(\theta) \approx \frac{1}{B}\sum_{i=1}^{B}\left[\left(\mathrm{Split}(Y_i, X_i) - b(X_i)\right)\nabla_\theta \log P_\theta(Y_i|X_i)\right]$$

Algorithm 1: NOFSS REINFORCE with Rollout Baseline

---

1 **Inputs:** θ, Number of epochs E, batch size B, number of instances K, number of clients n, vehicle capacity C, t-test threshold α

2 $T \leftarrow \dfrac{K}{B}$

3 $\theta^{BL} \leftarrow \theta$

4 **for** $e \leftarrow 1$ *to* E **do** // train for E epochs

5   **for** $t \leftarrow 1$ *to* T **do** // loop over the T instance batches

      // Get a batch of B CVRP instances with n clients

6        $X_i \leftarrow$ getInstance(n, C), $\forall i \in \{1, ..., B\}$

7

      // Sample a giant tour according to the learning policy P_θ

8        $Y_i \leftarrow$ SampleGiantTour(X_i, P_θ), $\forall i \in \{1, ..., B\}$

9

      // Generate a giant tour greedily according to the policy $P_{\theta BL}$

10       $Y_i^{BL} \leftarrow$ GreedyGiantTour$(X_i, P_{\theta BL})$, $\forall i \in \{1, ..., B\}$

11

      // Evaluate giant tours total travel cost

12       $L_i \leftarrow$ Split(X_i, Y_i, C) $\forall i \in \{1, ..., B\}$

13       $L_i^{BL} \leftarrow$ Split(X_i, Y_i^{BL}, C) $\forall i \in \{1, ..., B\}$

14

      // Compute the loss and update the neural network parameters

15       $\nabla_\theta \mathcal{L} \leftarrow \dfrac{1}{B} \sum_{i=1}^{B} (L_i - L_i^{BL}) \nabla_\theta \log P_\theta(Y_i | X_i)$

16       $\theta \leftarrow AdamW(\theta, \nabla_\theta \mathcal{L})$

17   **end**

18   **if** $t\text{-}test(P_\theta, P_{\theta BL}) < \alpha$ **then**

19       $\theta^{BL} \leftarrow \theta$

20   **end**

21 **end**

The baseline $b(X)$ is used to reduce the gradient variance, leading to an acceleration of the learning process. We use the greedy rollout baseline $b(X) =$ Split(Y^{BL}, X) which is an evaluation of the optimal Split of the giant tour Y^{BL} resulting from a copy of the learning neural network with parameters θ^{BL} that acts greedily, i.e. it chooses the next client with the highest probability of appearance at each time step. This baseline proved to be more efficient than actor-critic or REINFORCE with an exponential moving average baseline [5]. During validation, if the performance of θ is significantly better than that of θ^{BL} according to a t-test ($\alpha = 5\%$), the baseline is updated with the parameters of P_θ, i.e. θ^{BL} is set to θ.

4.1   Instance Features

For each instance X, we define the nodes and edges features as follows:

Node Features. Each node $u \in V$ is represented as a quadruplet $(x_u, y_u, \hat{d}_u, a_u)$ where (x_u, y_u) are the node coordinates sampled from a uniform distribution $\mathcal{U}([0,1] \times [0,1])$, $\hat{d}_u = d_u/C \in [0,1]$ is the normalized demand and $a_u = atan((y_u - y_0)/(x_u - x_0)) \in]-\pi/2, \pi/2[$ is the polar angle between the node u and the depot node 0.

Edge Features. For each edge $(u,v) \in E$, we define the edge features as the Euclidean distance between the nodes u and v (i.e. $d(u,v) := \|u - v\|, \forall (u,v) \in E$). The distance between two nodes in the instance is an interesting feature in the case of vehicle routing problems, since it is information that characterizes the problem well, and it appears in the objective function.

4.2   NOFSS Encoding-Decoding Architectures

The NOFSS approach is agnostic to the choice of the encoding and decoding model architectures. Thus, we propose to train various encoder-decoder models that rely on different graph neural networks (GNNs) and a GRU recurrent cell for decoding. The decoded sequence is passed to the Split algorithm in order to retrieve a candidate solution for the instance (Fig. 2).

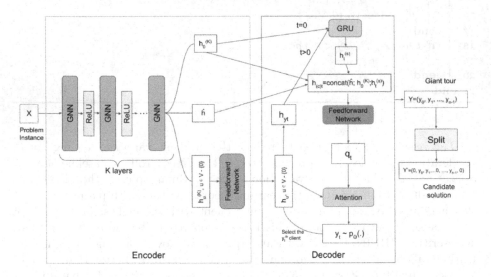

Fig. 2. Our proposed NOFSS model for solving CVRP instances.

Encoding. We experiment three GNN Encoders for our approach: GCN (a spectral GNN), GAT (a spatial GNN) and TransformerConv (a spatial GNN) [21]. Each encoder have K similar blocks. The GNN outputs an embedding for each node (clients and depot) $h_u^{(K)} \in \mathbb{R}^d, \forall u \in V$ and a graph representation computed using an average pooling $\bar{h} = 1/|V| \sum_{u \in V} h_u^{(K)}$. Finally, to distinguish the clients embeddings from the depot embedding $h_0^{(K)}$, we pass them into a feedforward layer $h_u = W_c \cdot h_u^{(K)} + b_c, \forall u \in V - \{0\}$, with $W_c \in \mathbb{R}^{d \times d}, b_c \in \mathbb{R}^d$ being respectively the weights and the bias of the layer.

Neighborhood Definition. As highlighted in Sect. 2, we can define a CVRP instance as a complete graph. We define the neighborhood $\mathcal{N}(u)$ of a client node $u \in V - \{0\}$ as the κ nearest nodes in terms of Euclidean distance and the depot 0, since it is important for the client's representation to be aware of the depot's existence (i.e. $\mathcal{N}(u) = \{v_1, v_2, ..., v_\kappa \in V; \|v_1 - u\| \leq \|v_2 - u\| \leq ... \leq \|v_\kappa - u\|\} \cup \{0\}$). For the depot, we consider that it is connected to every client. An example of an instance neighborhood definition is depicted in Fig. 3. The central node (red square) represents the depot, while the other nodes (blue circles) represent the clients. An edge exists between nodes u and v if $v \in \mathcal{N}(u)$. The number of nearest neighbors κ is determined per instance. We set it to be the average number of clients per route as if they were uniformly distributed on the routes, i.e. $\kappa = \dfrac{n}{m}$ with n being the number of clients and m being the lower bound of the number of routes. m is determined as the sum of all clients' demands divided by the vehicle's capacity rounded to the next integer $(m = \left\lceil \dfrac{\sum_{i=1}^{n} d_i}{C} \right\rceil)$. The advantage of such a definition of κ is that it takes into account the characteristics of the instance in terms of the number of clients, their demands, and the capacity of the vehicles instead of selecting an arbitrary number of neighbors.

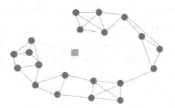

Fig. 3. CVRP instance with relationships between neighboring nodes (central square node is the depot).

Decoding. Since we are decoding a sequence of clients' order, we use a GRU recurrent cell [22]. GRU is relevant as it enables capturing the sequence representation while taking into account the order of its elements. It takes as input the previously selected client representation at step $t - 1$ concatenated with the depot representation $h_0^{(K)}$ and incorporates it in the global representation of the partial giant tour. At $t = 0$, we only use the depot representation $h_0^{(K)}$ as input to the GRU.

$$h_t^{(s)} = \begin{cases} \text{GRU}(h_0^{(K)}), & t = 0 \\ \text{GRU}([h_{y_{t-1}}; h_0^{(K)}]), & t > 0 \end{cases}$$

The graph embedding $\hat{h}$, the depot embedding $h_0^{(K)}$ and the sequence embedding $h_t^{(s)}$ are then concatenated together to form a context vector $h_c \in \mathbb{R}^{3d}$. The context vector is then passed to a feedforward layer made of two linear layers with $ReLU$ activation function in between to output a query vector $q_t \in \mathbb{R}^d$ i.e. $q = W_2 \cdot ReLU(W_1 \cdot h_c + b_1) + b_2$ with $W_2 \in \mathbb{R}^{d \times 3d}$, $W_2 \in \mathbb{R}^{d \times d}$, $b_1, b_2 \in \mathbb{R}^d$ being the parameters of the feedforward layer.

To compute the probability of selecting the next client $p_\theta(y_t | y_0, ..., y_{t-1}, X)$, we compute attention scores s_u ($\forall u \in V - \{0\}$) using a scaled dot-product with a masking mechanism in order to avoid selecting the same client twice. These scores are then clipped within $[-10, 10]$ using $tanh$ [5].

$$s_u = \begin{cases} c \cdot tanh\left(\dfrac{q_t h_u^\top}{\sqrt{d}}\right), & u \neq y_{t'}\ \ t' < t, c = 10 \\ -\infty & \text{otherwise} \end{cases}$$

The attention scores are converted into a probability distribution using the softmax function $p_i = p_\theta(y_t = i | y_0, ..., y_{t-1}, X) = softmax(s_i)$ By setting the value of the attention score to $-\infty$, we can perform the masking of already visited clients. Thus, when passed to the softmax function, its associated probability will be 0.

The Split Procedure. The algorithm works on the basis of the giant tour output by the neural network augmented with the depot, i.e. $\mathcal{Y} = (y_0, y_1, ..., y_n)$ with $y_0 = 0$ being the depot. Using the giant tour, we define an auxiliary graph $H(V^H, E^H)$ with $|V^H| = n + 1$. The nodes in V^H indicate the depot (either for return or departure). The edge set indicates all possible sub-sequences that starts from y_i to y_j ($y_i, y_{i+1}, ..., y_j$) that do not transgress the vehicle's capacity constraint. We formulate it as follows: $E^H = \{(i, j) \in V^H \times V^H;\ \ i < j,\ \ \sum_{k=i+1}^{j} d_{y_k} \leq C\}$. The edges are weighted as follows: for an edge $(i, j) \in E^H$ we associate the total travelled distance starting from the depot to the client y_{i+1}, visiting the tour $(y_{i+1}, ..., y_j)$ and going back to the depot from y_j:

$$D^H = \{d_{ij} = dist(0, y_{i+1}) + \sum_{\substack{k=i+1 \\ j-i>1}}^{j-1} dist(y_k, y_{k+1}) + dist(y_j, 0), \quad \forall(i, j) \in E^H\}$$

This gives us a direct acyclic graph where we solve a shortest path problem using Bellman's algorithm. The associated shortest path cost represents the best solution length (total travelled distance) for the CVRP instance with regard to the given giant tour.

5   Experiments

Data Generation. We follow the data generation protocol of Nazari et al. [6] to consider 3 types of CVRP instances with number of clients $n = 20$, 50 and 100. For each problem size, we have generated $100k$ instances for training, and two sets of $10k$ instances for validation and test. Clients and depot locations are generated from a uniform distribution $\mathcal{U}(\{[0,1] \times [0,1]\})$. The clients' demands are also uniformly drawn from the interval $[1,9]$. Vehicles' capacities are set to 30, 40 and 50 respectively for $n = 20$, 50, 100.

Hyperparameters. We use an embedding dimension $d = 128$ and a uniform parameter initialization for our deep neural networks $\mathcal{U}(-1/\sqrt{d}, 1/\sqrt{d})$ and set the learning rate to $\eta = 10^{-3}$. The models are trained with a time limit of 100 hours and batch size $B = 128$ on a single NVIDIA V100 GPU with 16 GB of VRAM. For each encoder type, we use $K = 3$ GNN blocks. Implementations use PyTorch and PyTorch Geometric for graph neural networks [23] (Python), while the Split algorithm is implemented in C.

Baselines. We use HGS[4] [16] as baseline as it is one of the state of the art metaheuristics for the CVRP. We also use classical CVRP heuristics[5]: (i) RFCS [7] as a two-step order-first split-second heuristic, (ii) Sweep [12] as a two-step cluster-first route-second approach, and (iii) Nearest Neighbor heuristic as a single-step construction approach [24]. We also trained the model with TransformerConv encoder in an end-to-end manner for depot and clients choice (Full-learning). We first note that NOFSS models are faster to train, completing $E = 1000$ of learning epochs in the 100 h time budget, while the Full-learning models perform 1000, 500 and 200 training epochs for instance sizes of 20, 50 and 100 respectively. For the exploitation of the learned policies, we use a greedy decoding which considers the highest probability at each decoding step and a sampling strategy which samples 1280 candidate solutions for each test instance from the probability distributions given by the models. Table 1 reports the results of each approach on the test specifying: average solution lengths (obj.), the average gap (in percentage) to the best average solution lengths and the running time (in seconds) to output a candidate solution for a single instance.

5.1   Comparison with a Full-Learning Setting

Figure 4 presents the evolution of the average solution length per epoch during training and validation on CVRP instances with 20 clients (left) and 50 clients (right). During training, candidate solutions are sampled from the model and their total lengths are averaged over the training set. Let us note that the models' parameters are updated each time a batch is processed via gradient descent,

---

[4] https://github.com/vidalt/HGS-CVRP.
[5] https://github.com/yorak/VeRyPy.

Table 1. NOFSS vs. other algorithms. FL for Full-Learning; exploitation, greedy (G), sampling (S).

Method	$n = 20$			$n = 50$			$n = 100$		
	obj	gap (%)	time (s)	obj	gap (%)	time (s)	obj	gap (%)	time (s)
HGS	6.13	0.00	0.003	10.34	0.00	0.09	15.57	0.00	0.69
RFCS	6.30	2.76	0.02	10.90	5.39	0.57	16.62	6.73	7.53
Sweep	7.55	23.16	0.01	15.60	50.93	0.06	28.56	83.37	0.23
Nearest neighbor	7.39	20.57	0.0004	12.63	22.19	0.001	18.95	21.68	0.01
NOFSS-GCN (G)	6.83	11.41	0.0008	12.31	19.05	0.003	19.41	24.66	0.007
NOFSS-GAT (G)	6.59	7.50	0.006	11.74	13.53	0.02	18.34	17.80	0.05
NOFSS-Transformer (G)	6.50	6.03	0.006	11.57	11.89	0.02	18.13	16.44	0.06
FL-Transformer (G)	6.49	5.87	0.006	11.34	9.67	0.02	17.69	13.61	0.06
NOFSS-Transformer (S)	6.24	1.79	1.37	11.03	6.67	1.56	17.45	12.07	2.43
FL-Transformer (S)	6.18	0.81	2.09	10.79	4.35	2.35	17.32	11.23	8.29

thus the performance of the models changes every batch during training, while validation is performed using the model resulting from the processing of the last batch in the training set, which is theoretically the best model achieved at the end of the epoch. Also, in validation, we use a greedy decoding instead of sampling. The evolution of the average solution lengths shows that the NOFSS model is able to learn an implicit policy for solving the CVRP by learning to output an indirect representation of the solution. On instances with 20 clients, we can observe that during training, the NOFSS model achieves better average solution lengths than the Full-learning model. On validation, we observe the same trend as in training, but starting from the 600th epoch, the Full-learning model slightly outperforms the NOFSS model. The equivalent performance of the two models is confirmed on the test set with average solution lengths of 6.50 and 6.49 on greedy decoding for NOFSS and Full-learning respectively with similar execution times. On sampling decoding, similar performances are observed, with 0.9% difference in performance between the two models, but with an advantage in execution time in favor of NOFSS. On CVRP with 50 clients, we observe that NOFSS has a better jump start performance on training and a better final performance for the Full-learning model. We observe 2% difference in performance for greedy and sampling decoding on the test set. We also note similar sampling times for the two types of models in greedy decoding, while NOFSS being 52%, 50% and 241% faster in sampling respectively for $n = 20$, 50 and 100.

5.2   Comparison to Handcrafted Heuristics

When compared to handcrafted heuristics, we can observe from Table 1 that either with greedy or sampling exploitation, NOFSS models outperform the Sweep and Nearest neighbor algorithms. NOFSS model seems to output better solution lengths, on average, than RFCS on CVRP with 20 clients when using the sampling strategy but seems to fail scaling to CVRP with 50 and 100

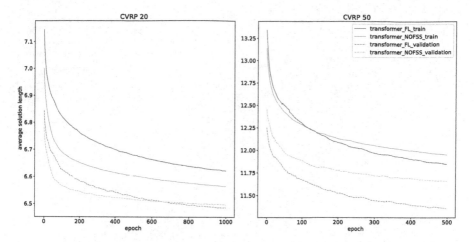

Fig. 4. Learning curves in training and validation for Full-learning (blue) and NOFSS models (orange) on CVRP instances with 20 (CVRP20) and 50 clients (CVRP50); lower is better. (Color figure online)

clients. Let us note that while RFCS and NOFSS belong to the same type of two-step strategy, there is a difference in the two approaches in that RFCS explicitly solves a Traveling Salesman Problem, while NOFSS directly evaluates the giant tour using the Split algorithm. The difference in average solution lengths may suggest that NOFSS learned policy is different from a policy that learns to solve a Traveling Salesman Problem.

5.3 Influence of the Type of Encoder

We investigate the influence of the choice of GNN encoder on models' performance. Figure 5 shows the evolution of the average solutions lengths per epoch in training and validation phases for the 3 types of GNN encoders: GCN, GAT and TransformerConv on CVRP with 20 and 50 clients. We observe the same trends for both training and validation phases, with TransformerConv having the best convergence, followed by GAT encoder and finally by GCN encoder. The instances' representation plays an important role in the resolution process, because a good representation leads to the exploitation of meaningful features and, thus, gives a better solution. The choice of the encoder seems to be a critical part of the model's architecture. It appears from these results that spatial GNNs better perform than spectral GNNs in our evaluation setting. Exploiting the graph topology in the spatial domain seems to benefit more in the context of vehicle routing problems than exploiting the graph structure in the spectral domain. While TransformerConv and GAT are both spatial GNNs, it seems that the way they exploit the node and edges information has an impact on the overall performance of the models.

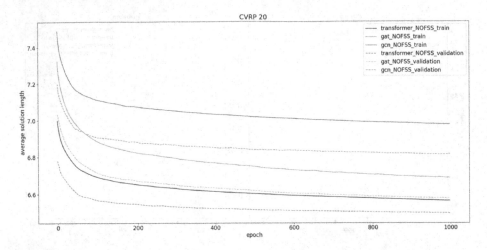

Fig. 5. Comparison of Graph Neural Network encoders on models' performance (training and validation).

5.4   On Models Generalization

We propose to study the generalization of the models trained on a set of instances with a specific size to instances of different size. For this, we evaluate the different test sets on instances of different sizes. For example, we evaluate the NOFSS Transformer model trained on CVRP with 20 clients instances (Transformer-20) on instances with 20, 50 and 100 clients. Table 2 sums up our results. We report the average solution lengths for both greedy and sampling exploitation strategies. For greedy decoding, we report the results for the models trained on the different instance sizes while for sampling, we focus on the model trained on instances sizes which seems more promising based on our findings on the greedy decoding. We observe that for the Transformer-20, the NOFSS model has a better generalization property than the Full-learning model, with performance similar for $n = 20$ and $n = 100$ and better for $n = 50$. Since training models on instances with 20 clients is faster, it is relevant to identify that the NOFSS model is a better choice.

For Transformer-50 and Transformer-100, it appears that, for $n = 20$ NOFSS models have better performances than their Full-learning counterparts while staying competitive for $n = 50$ and $n = 100$. An interesting result observed on Transformer-50 is its good generalization to CVRP instances with 100 clients, as it appears that it achieves better performance than the models trained on instances with 100 clients. This may suggest that relevant invariants that are beyond the instance size are learned while training on instances with 50 clients. We push further our investigations on Transformer-50 by analyzing its performance with a sampling exploitation strategy. While for the instances with 20 clients, the models stay competitive with the ones trained on that size, they achieve the best performances on the sets with instances with 50 and 100 clients.

Transformer-50 appears to be a good trade-off between learning speed (it is faster to train than Transformer-100) and performance.

Table 2. Comparison of average solution lengths achieved by the NOFSS and Full-learning models on different instance sizes of the test set.

Trained model	NOFSS (G)			Full-learning (G)		
	20	50	100	20	50	100
Transformer-20	6.50	11.62	18.34	6.49	12.01	18.33
Transformer-50	6.64	11.57	17.97	6.76	11.34	17.52
Transformer-100	6.94	11.79	18.13	6.98	11.65	17.69
	NOFSS (S)			Full-learning (S)		
	20	50	100	20	50	100
Transformer-50	6.31	11.03	17.40	6.25	10.79	17.22

6 Conclusion

In this work, we proposed NOFSS, a two-step algorithm hybridizing a deep neural network model and an exact tour splitting procedure for the Capacitated Vehicle Routing Problem. To the best of our knowledge, this is the first model that proposes a hybridization between a deep neural network and a dynamic programming algorithm to successfully learn an implicit policy based on giant tour generation to solve the CVRP. We conducted extensive experiments on the proposed models with various Graph Neural Network encoders and compared them against classic CVRP heuristics and an end-to-end Full-learning model. Our results show that NOFSS is very competitive, even if it currently does not surpass end-to-end full-learning approaches. NOFSS is however faster than end-to-end approaches in both training and evaluation. It also shows good generalization properties when trained on instances with a specific size and applied to solve instances of different sizes. The NOFSS model is easier to implement than an end-to-end learning-based policy and does not rely on sophisticated handcrafted search strategies to find good quality solutions.

Future work should investigate more on the generalization of the method to instances of bigger sizes. Also, while we tested only greedy and sampling strategies for exploiting the trained models, other relevant strategies may be interesting such as beam search, or using bigger sample sizes than the one we used since NOFSS has a faster execution time. The solution given by NOFSS can also be a good warm start for further improvement by local search algorithms. Finally, since our approach is generic, it would be interesting to evaluate it on other problems, such as the Vehicle Routing Problem with Time Windows.

References

1. Bengio, Y., Lodi, A., Prouvost, A.: Machine learning for combinatorial optimization: a methodological tour d'horizon. Eur. J. Oper. Res. **290**, 405–421 (2021)
2. Vinyals, O., Fortunato, M., Jaitly, N.: Pointer networks. arXiv:1506.03134 (2015)
3. Bello, I., Pham, H., Le, Q.V., Norouzi, M., Bengio, S.: Neural combinatorial optimization with reinforcement learning. arXiv:1611.09940 (2016)
4. Deudon, M., Cournut, P., Lacoste, A., Adulyasak, Y., Rousseau, L.-M.: Learning heuristics for the TSP by policy gradient. In: van Hoeve, W.-J. (ed.) CPAIOR 2018. LNCS, vol. 10848, pp. 170–181. Springer, Cham (2018). https://doi.org/10.1007/978-3-319-93031-2_12
5. Kool, W., Van Hoof, H., Welling, M.: Attention, learn to solve routing problems! arXiv:1803.08475 (2018)
6. Nazari, M., Oroojlooy, A., Snyder, L.V., Takáč, M.: Reinforcement learning for solving the vehicle routing problem. arXiv:1802.04240 (2018)
7. Beasley, J.E.: Route first-cluster second methods for vehicle routing. Omega **11**, 403–408 (1983)
8. Prins, C.: A simple and effective evolutionary algorithm for the vehicle routing problem. Comput. Oper. Res. **31**, 1985–2002 (2004)
9. Toth, P., Vigo, D.: The Vehicle Routing Problem. SIAM, Philadelphia (2002)
10. Smith, K.A.: Neural networks for combinatorial optimization: a review of more than a decade of research. INFORMS J. Comput. **11**, 15–34 (1999)
11. Vaswani, A., et al.: Attention is all you need. In: Advances in Neural Information Processing Systems, pp. 5998–6008 (2017)
12. Gillett, B.E., Miller, L.R.: A heuristic algorithm for the vehicle-dispatch problem. Oper. Res. **22**, 340–349 (1974)
13. Ryan, D.M., Hjorring, C., Glover, F.: Extensions of the petal method for vehicle routeing. J. Oper. Res. Soc. **44**, 289–296 (1993)
14. Fisher, M.L., Jaikumar, R.: A generalized assignment heuristic for vehicle routing. Networks **11**, 109–124 (1981)
15. Hiquebran, D., Alfa, A., Shapiro, J., Gittoes, D.: A revised simulated annealing and cluster-first route-second algorithm applied to the vehicle routing problem. Eng. Optim. **22**, 77–107 (1993)
16. Vidal, T.: Hybrid genetic search for the CVRP: open-source implementation and swap* neighborhood. Comput. Oper. Res. **140**, 105643 (2022)
17. Prins, C., Lacomme, P., Prodhon, C.: Order-first split-second methods for vehicle routing problems: a review. Transp. Res. Part C: Emerg. Technol. **40**, 179–200 (2014)
18. Veličković, P., Cucurull, G., Casanova, A., Romero, A., Lio, P., Bengio, Y.: Graph attention networks. arXiv:1710.10903 (2017)
19. Kipf, T.N., Welling, M.: Semi-supervised classification with graph convolutional networks. arXiv:1609.02907 (2016)
20. Zhou, J., et al.: Graph neural networks: a review of methods and applications. AI Open **1**, 57–81 (2020)
21. Shi, Y., Huang, Z., Feng, S., Zhong, H., Wang, W., Sun, Y.: Masked label prediction: unified message passing model for semi-supervised classification. arXiv:2009.03509 (2020)
22. Cho, K., Van Merriënboer, B., Bahdanau, D., Bengio, Y.: On the properties of neural machine translation: Encoder-decoder approaches. preprint arXiv:1409.1259 (2014)

23. Fey, M., Lenssen, J.E.: Fast graph representation learning with PyTorch geometric. In: ICLR Workshop on Representation Learning on Graphs and Manifolds (2019)
24. Rasku, J., Kärkkäinen, T., Musliu, N.: Meta-survey and implementations of classical capacitated vehicle routing heuristics with reproduced results. Toward Automatic Customization of Vehicle Routing Systems (2019)

Applications

GRASP-Based Hybrid Search to Solve the Multi-objective Requirements Selection Problem

Víctor Pérez-Piqueras[✉] [iD], Pablo Bermejo López[iD], and José A. Gámez[iD]

Department of Computing Systems, Intelligent Systems and Data Mining Laboratory (I3A), Universidad de Castilla-La Mancha, 02071 Albacete, Spain
{victor.perezpiqueras,pablo.bermejo,jose.gamez}@uclm.es

Abstract. One of the most important and recurring issues that the development of a software product faces is the requirements selection problem. Addressing this issue is especially crucial if agile methodologies are used. The requirements selection problem, also called Next Release Problem (NRP), seeks to choose a subset of requirements which will be implemented in the next increment of the product. They must maximize clients satisfaction and minimize the cost or effort of implementation. This is a combinatorial optimization problem studied in the area of Search-Based Software Engineering. In this work, the performance of a basic genetic algorithm and a widely used multi-objective genetic algorithm (NSGA-II) have been compared against a multi-objective version of a randomized greedy algorithm (GRASP). The results obtained show that, while NSGA-II is frequently used to solve this problem, faster algorithms, such as GRASP, can return solutions of similar or even better quality using the proper configurations and search techniques. The repository with the code and analysis used in this study is made available to those interested via GitHub.

Keywords: GRASP · Multi-objective optimization · Next release problem · Requirements selection · Search-based software engineering

1 Introduction

Software systems are increasing in functionality and complexity over time. This implies that new software projects are potentially more complicated to manage and complete successfully. One of the problematics that can heavily affect the outcome of a project is the planning of a release. In a software project, the product to be delivered is defined by a set of software requirements. These requirements are offered to a group of clients, who will give feedback on which requirements are more important to them. Then, a set of requirements is planned

This work has been partially funded by the Regional Government (JCCM) and ERDF funds through the projects SBPLY/17/180501/000493 and SBPLY/21/180501/000148.

for the release. Selecting the requirements that better fit client interests starts getting complicated when development capacity has to be taken into account. Furthermore, requirements can have dependencies between them. This problem, named requirements selection problem, is very complex and does not have a unique and optimal solution. Two objectives coexist: maximizing the satisfaction of the clients and minimizing the effort of the software developers. Therefore, solutions can range from sets of few requirements with minimal effort and satisfaction, to sets of plenty of requirements, which will imply high client satisfaction but at the cost of a high effort.

Thus, this planning step is critical, especially when applying incremental software development methodologies due to the need to solve the requirements selection problem multiple times, at each iteration. Thus, this problem is candidate to be automated by means of optimization methods. Previous works have studied the applicability of different search techniques, giving preference to evolutionary algorithms, mainly. In our study, we present an algorithm based on the Greedy Randomized Adaptive Search Procedure (GRASP, [7]). We have explored new procedures that allow to improve GRASP performance in the requirements selection problem beyond that of previous studies. The experimentation that we carried out shows that the GRASP metaheuristic can obtain similar results as those of the evolutionary approaches, but reducing drastically its computational cost.

The rest of the paper is structured as follows. In Sect. 2, a summary of previous works and procedures they applied is made. Section 3 describes our algorithm proposal and defines the solution encoding along with the most important methods and techniques. Then, in Sect. 4, the evaluation setup is described, along with the algorithms, datasets and methodology used. Section 5 presents and discusses the results of the experimentation. Finally, Sect. 6 summarizes the conclusions of this study and introduces potential new lines of work for the future.

2 Requirements Selection

2.1 Related Work

The requirements selection problem is studied in the Search-Based Software Engineering (SBSE) research field, where Software Engineering related problems are tackled by means of search-based optimization algorithms. The first definition of the requirements selection problem was formulated by Bagnall et al. [1]. In their definition of the Next Release Problem (NRP), a subset of requirements has to be selected, having as goal meeting the clients[1] needs, minimizing development effort and maximizing clients satisfaction. In their work, different metaheuristics algorithms, such as simulated annealing, hill climbing and GRASP algorithms were proposed, but all of them combined the objectives of the problem using an aggregate function. The same procedure of single-objective

---

[1] Although "stakeholder" is a more appropriate term, "client" will be used to keep coherence with previous works present in the literature.

proposals was followed by Greer and Ruhe [9]. They studied the generation of feasible assignments of requirements to increments, taking into account different resources constraints and stakeholders perspectives. Genetic algorithms (GAs) were the optimization technique selected to solve the NRP. Later, Baker et al. [2] demonstrated that metaheuristics techniques could be applied to real-world NRP outperforming expert judgement, using in their study simmulated annealing and greedy algorithms. The works of del Sagrado et al. [5] applied ACO (Ant Colony Optimization). All of these approaches followed a single-objective formulation of the problem, in which the aggregation of the objectives resulted in a biased search.

It was not until the proposal of Zhang et al. [13] that the NRP was formulated as a multi-objective optimization (MOO) problem. This new formulation, Multi-Objective Next Release Problem (MONRP), formally defined in Sect. 2.2, was based on Pareto dominance. Their proposal tackled each objective separately, exploring the non-dominated solutions. Finkelstein et al. [8] also applied multi-objective optimization considering different measures of fairness. All these studies applied evolutionary algorithms, such as ParetoGA and NSGA-II [4] to solve the MONRP.

Other works that kept exploring evolutionary algorithms to solve the MONRP are those of Durillo et al. [6]. They proposed two GAs, NSGA-II and MOCell (MultiObjective Cellular genetic algorithm), and an evolutionary procedure, PAES (Pareto Archived Evolution Strategy).

2.2 Multi-objective Formulation

As mentioned in the introduction, the NRP requires a combinatorial optimization of two objectives. While some studies alleviate this problem by adding an aggregate (single-objective optimization), others tackle the two objectives by using a Pareto front of non-dominated solutions (MOO). Defining the NRP as a multi-objective optimization problem gives the advantage that a single solution to the problem is not sought, but rather a set of non-dominated solutions. In this way, one solution or another from this set can be chosen according to the conditions, situation and restrictions of the software product development. This new formulation of the problem is known as MONRP.

The MONRP can be defined by a set $R = \{r_1, r_2, \ldots, r_n\}$ of n candidate software requirements, which are suggested by a set $C = \{c_1, c_2, \ldots, c_m\}$ of m clients. In addition, a vector of costs or efforts is defined for the requirements in R, denoted $E = \{e_1, e_2, \ldots, e_n\}$, in which each e_i is associated with a requirement r_i. Each client has an associated weight, which measures its importance. Let $W = \{w_1, w_2, \ldots, w_m\}$ be the set of client weights. Moreover, each client gives an importance value to each requirement, depending on the needs and goals that this has with respect to the software product being developed. Thus, the importance that a requirement r_j has for a client c_i is given by a value v_{ij}, in which a zero value represents that the client c_i does not have any interest in the implementation of the requirement r_j. A $m \times n$ matrix is used to hold all the importance values in v_{ij}. The overall satisfaction provided by a requirement

r_j is denoted as $S = \{s_1, s_2, \ldots, s_n\}$ and is measured as a weighted sum of all importance values for all clients. The MONRP consists of finding a decision vector X, that includes the requirements to be implemented for the next software release. X is a subset of R, which contains the requirements that maximize clients satisfaction and minimize development efforts.

3   Proposal

Evolutionary algorithms have been widely applied to solve the MONRP [3,11,13]. Most of the previous studies are based on algorithms such as NSGA-II, ParetoGA or ACO, usually comparing their performance with other algorithms less suited to the problem, e.g. generic versions of genetic or greedy algorithms. However, evolutionary approaches involve a high computational cost, as the algorithms have to generate a population of solutions and evolve each one of the solutions applying many operators and fitness evaluations. For this reason, in this work we have pursued to design an algorithm that can return a set of solutions of similar quality to those obtained by the evolutionary proposals, but reducing the cost of its computation. In this section, it is presented the multi-objective version of a greedy algorithm, along with the solution encoding used. Then, the most relevant procedures of the algorithm and enhancements are presented.

3.1   GPPR: A GRASP Algorithm with Pareto Front and Path Relinking

GRASP is a multi-start method designed to solve hard combinatorial optimization problems, such as the MONRP. It has been used in its simplest version [3,11] to solve the requirements selection problem.

The basic actions of a canonical GRASP procedure consist of generating solutions iteratively in two phases: a greedy randomized construction and an improvement by local search (see Sects. 3.3 and 3.4). These two phases have to be implemented specifically depending on the problem at hand.

We have designed a variant of the GRASP procedure with the goal of solving the MONRP in a hybrid manner, that is, applying both single-objective and multi-objective search methods. The algorithm, named GRASP algorithm with Pareto front and Path Relinking (GPPR), executes a fixed number of iterations, generating at each iteration a set of solutions, instead of only one solution per iteration (which is a different approach from the canonical GRASP that generates one solution per iteration, but in the end works identically). Additionally, we have extended the procedure, updating the Pareto front with the new solutions found after each iteration, and adding a post-improvement procedure known as Path Relinking (see Sect. 3.5), that will enhance the quality of the Pareto front found by exploring trajectories that lead to new non-dominated solutions. The pseudocode of GPPR is shown in Algorithm 1.

Each one of the operators included in the pseudocode is described in detail in Sects. 3.3, 3.4 and 3.5, respectively. As explained previously, GPPR is an

---

Algorithm 1. GPPR pseudocode

---

  procedure GPPR($maxIterations$)
      $nds \leftarrow \emptyset$ ▷ empty set of non-dominated solutions
      for $i = 0$ **to** $maxIterations$ **do**
         $solutions \leftarrow$ constructSolutions()
         $solutions \leftarrow$ localSearch($solutions$)
         $solutions \leftarrow$ pathRelinking($solutions, nds$)
         $nds \leftarrow$ updateNDS($solutions$)
      end for
      return nds
  end procedure

---

algorithm that applies a hybrid approach. It maintains and updates at each iteration a set of non-dominated solutions, returned in the form of a Pareto front at the end of the search. However, it uses an aggregate of the two problem objectives in some phases of the execution (depending on the methods chosen for each phase).

3.2 Solution Encoding

Each candidate solution in GPPR is represented by a vector of booleans of length n. Each value of the vector indicates the inclusion or not of a requirement of the set R (see Sect. 2.2). The satisfaction and effort of each requirement are scaled using a min-max normalization. Each solution is evaluated by means of a *singleScore* value that mixes the scaled satisfaction and effort of the set X of selected requirements in the solution . In this version of the GPPR, we did not model cost restrictions nor interactions between requirements .

3.3 Construction

In this phase a number of solutions are constructed. Their generation can be either randomized or stochastic. We have designed two methods for the construction phase:

- **Uniform.** First, the number x of selected requirements is randomly chosen. Then, x requirements are selected randomly, having each requirement r of the set R of length n a probability $\frac{1}{n}$ of being selected. This construction method works as a random selection of requirements.
- **Stochastic.** The probability of each requirement being selected is proportional to its *singleScore*.

3.4 Local Search

This phase is executed after the construction of an initial set of solutions, and it aims to find solutions in the neighbourhood that enhance the former ones.

Since GPPR aims to generate solutions fast, it performs a ranking-based forward search, in which it tries to find and return a neighbour that is better than the initial one. This search method tends to fall into local optima, but it can be corrected increasing the number of executions or applying extra operators after this phase (see Subsect. 3.5).

3.5   Path Relinking

One of the adverse characteristics of GRASP is its lack of memory structures. Iterations in GRASP are independent and do not use previous observations. Path Relinking (PR) is a possible solution to address this issue. PR was originally proposed as a way to explore trajectories between elite solutions. In the problem being tackled, elite solutions are the non-dominated solutions. Using one or more elite solutions, trajectories that lead to other elite solutions in the search space are explored, in order to find better solutions. PR applied to GRASP was introduced by Laguna and Martí [10]. It has been used as an intensification scheme, in which the generated solutions of an iteration are relinked to one or more elite solutions, creating a post-optimization phase.

The PR method can be applied either after each iteration, involving a higher computational cost; or at the end of the execution, relinking only the final elite solutions, reducing the effectiveness of this method but speeding up the execution.

In this proposal we have decided to apply PR at each iteration as a third phase, after the local search. The pseudocode is described in Algorithm 2. For each one of the solutions found after the local search, this procedure will try to find a path from each solution to a random elite solution from the set of non-dominated solutions (NDS). This path will help the procedure to find intermediate solutions that can possibly be better than the former ones. For this purpose, each solution in the current set of solutions obtained after the local search is assigned an elite solution from the current NDS. Then, it calculates the Hamming distance of these two solutions. Having the distance value, the procedure finds the bits that are different, that is, the requirements included in one solution that are not in the other. Then, it updates the current solution (flips the bit that returns the highest *singleScore* value) and saves the new path solution in a solution path list, decrementing the distance from the current solution to the elite one. When the distance is zero, the best solution found in the path is appended to a set of best solutions (*bestSols* in Algorithm 2) found by the PR procedure. After finding best path solutions for all the initial solutions, the procedure returns the set of former solutions plus the new solutions found.

4   Evaluation Setup

In this section, we present the experimental evaluation. We describe competing approaches used to be compared against our proposal, along with the datasets used to evaluate the algorithms. Our algorithms have been implemented in

---

Algorithm 2. Path Relinking pseudocode

---

procedure PATHRELINKING(*solutions*, *nds*)
    bestSols ← ∅
    for *sol* in *solutions* **do**
        currSol ← *sol*   ▷ Create a copy to be modified
        eliteSol ← getRandomSol(*nds*)
        distance ← countDistance(*currSol*, *eliteSol*)
        pathSols ← ∅
        while *distance* > 0 **do**
            diffBits ← findDiffBits(*currSol*, *eliteSol*)
            currSol ← flipBestBitSingleScore(*diffBits*)
            pathSols ← savePath(*currSol*)
            distance ← *distance* − 1
        end while
        bestSols ← *bestSols* ∪ findBestSol(*pathSols*)
    end for
    return *solutions* ∪ *bestSols*
end procedure

Python 3.8.8. The source code, experimentation setup and datasets are available at the following repository: https://github.com/UCLM-SIMD/MONRP/tree/ola22.

4.1 Algorithms

To properly compare the effectivity and performance of our proposal, besides GPPR we have included in our experiments the following algorithms: Random search, Single-Objective GA and NSGA-II. The ranges of parameters used in the experimentation for each algorithm are described in Sect. 4.3, along with their descriptions.

4.2 Datasets

We have tested the performance of the algorithms using a variety of datasets from different sources. Datasets P1 [9] and P2 [11] include 5 clients and 20 requirements, and 5 clients and 100 requirements, respectively.

Due to the privacy policies followed by software development companies, there is a lack of datasets to experiment with. For this reason, we have created synthetically a larger dataset (S3) that includes 100 clients and 140 requirements, in order to evaluate the shift in performance of the algorithms.

4.3 Methodology

We tested a set of configurations for each algorithm and dataset. Each configuration was executed 10 times. For the Single-Objective GA and NSGA-II, populations were given values among $\{20, 30, 40, 100, 200\}$ and number of generations took values $\{100, 200, 300, 500, 1000, 2000\}$. Crossover probabilities range from $\{0.6, 0.8, 0.85, 0.9\}$. Two mutation schemes were used, *flip1bit* and *flipeach-bit*, and mutation probabilites from $\{0, 0.05, 0.1, 0.2, 0.5, 0.7, 1\}$. Both algorithms used a binary tournament selection and a one-point crossover scheme. For the replacement scheme, both Single-Objective GA and NSGA-II applied elitism. The total amount of different hyperparameter configurations executed for each GA and each dataset was 1680. Our GPPR algorithm was tested using a number of iterations from $\{20, 40, 60, 80, 100, 200, 500\}$ and a number of solutions per iteration from $\{20, 50, 100, 200, 300, 500\}$. We tested all combinations of construction methods, local search and PR (including configurations with no local search and no PR methods), which resulted in 1008 different hyperparameter configurations executed for each dataset.

The stop criterion used in other works [3,11,13] is the number of function evaluations, commonly set to 10000. To adapt our experiments to this stop criterion, we restricted the execution of our GAs to: *Pop. size* $\times$ *#Gens.* ≤ 10000; and for the GPPR: *Iterations* $\times$ *Sols. per Iteration* ≤ 10000.

The GPPR normalizes the satisfaction and effort values, scaling them between 0 and 1. To properly compare its Pareto front solutions against those returned by the GAs, these evolutionary approaches have also used the normalized version of the dataset values. To evaluate the results, we compared the obtained Pareto fronts and a set of quality indicators of the results generated by the algorithms and their efficiency:

- **Hypervolume (HV).** Denotes the space covered by the set of non-dominated solutions [14]. Pareto fronts with higher HV are preferred.
- **Δ-Spread.** It measures the extent of spread achieved among the obtained solutions [6]. Pareto fronts with lower Δ-Spread are preferred.
- **Spacing.** It measures the uniformity of the distribution of non-dominated solutions [12]. Pareto fronts with greater spacing are preferred.
- **Execution time.** The total time taken by the algorithm to finish its execution. Algorithms with lower execution time are preferred.

Mean values of these metrics have been calculated and compared in a pairwised manner between algorithms using the Wilcoxon rank-sum non-parametric test, which allows to assess whether one of two samples of independent observations tends to have larger values than the other.

5   Results and Analysis

5.1   Best Configurations

The Single-Objective GA's best hyperparameter configuration includes a population size of 100 individuals, a number of generations of 100 (maximum number to stay under the 10,000 limit) and a $P_c = 0.8$. The mutation operator that showed a better performance was the *flip1bit*. This operator gives a chance of

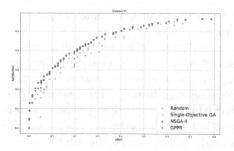

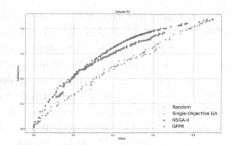

Fig. 1. Pareto front for dataset P1   **Fig. 2.** Pareto front for dataset P2

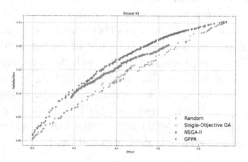

Fig. 3. Pareto front for dataset S3

flipping only one bit of the booleans vector. The best-performing probability is $P_m = 1$, which means that we always mutates one random bit of each individual . That probability is equivalent to using $P_m = \frac{1}{n}$ at gene level, n being the number of genes (scheme used in [6, 11]). The best hyperparameter configuration for the NSGA-II used a population size of 100 individuals and 100 generations. The best crossover probability (P_c) was the lowest, 0.6, and the best mutation operator was the *flip1bit*, using a $P_m = 1$. For the GPPR , the ratio between iterations and number of solutions per iteration is less important, as this algorithm does not have memory. Thus, a similar hyperparameter configuration to those of the GAs was used. The construction method that showed a better performance was the *stochastic* one, giving preference to requirements with higher *singleScore*. In all scenarios, hyperparameter configurations with *uniform* construction performed worse.

5.2   Pareto Results

Pareto results are shown in Figs. 1, 2 and 3. For the sake of space, we omit results of worst-performing hyperparameter configurations.

The Single-Objective GA shows bad performance, being similar to that of the random procedure. This occurs due to the low number of generations set to keep the maximum number of function evaluations. Configuring a number of generations of one or two magnitude orders higher increases the quality of its

Pareto front. Regarding the Pareto front distribution, this GA's aggregation of objectives biases the search, leaving unexplored areas.

The NSGA-II algorithm generates Pareto fronts of better quality: better solutions and more distributed along the search space. As expected, the crowding operator of the algorithm helps exploring the search space. However, as the dataset size increases, its performance decreases significantly. The reason is the limited number of generations, as this algorithm is expected to perform better in larger datasets when compared against other search methods.

Multi-Objective (MO) local search methods do not worsen the solutions, but do not improve them either, as a GPPR without local search is returning similar Pareto fronts. This implies that local search methods that can only explore the neighbourhood of a solution are not able to improve the solutions. Nevertheless, the PR procedure is capable of finding better solutions. In all cases, algorithms applying this methods returned Pareto fronts of higher quality. Regarding the Pareto distribution, as this procedure starts each iteration randomly, it explores the majority of the search space. The most interesting feature of the GPPR is that, while the dataset size grows, its performance is not demeaned. Therefore, unless the search space is much larger, our GPPR proposal can return a Pareto front of acceptable quality very efficiently, while GAs require higher number of iterations, impplying a less affordable computational cost.

5.3   Metrics Results

The mean values of the metrics obtained for each algorithm and dataset after 10 independent runs have been statistically compared, as explained in Sect. 4.3. Each metric mean value has been compared pair-wise between algorithms, denoting the best value in **bold** and indicating the values that are statistically worse ($P < 0.05$) with a $\downarrow$ symbol (see Table 1).

Regarding the HV metric, our GPPR algorithm has obtained significantly better results than the two GAs in datasets P2 and S3, and worse results for dataset P1, whose search space is very small. These results only denote that the extreme solutions of the Pareto front returned by GPPR cover a larger area than those obtained by the GAs.

The Δ-Spread values show that the Single-Objective GA obtains the lowest values, that is, the best Δ-Spread values. Nevertheless, our GPPR proposal obtains values lower than those of the NSGA-II, outperforming it once again.

The spacing values show that, again, the GPPR outperforms the two GAs in the two larger datasets. Comparing the spacing values of the smallest dataset, P1, it is observed that GPPR spacing values decrease, being close to those obtained by the NSGA-II algorithm. However, in datasets P2 and S3, the GPPR spacing values are significantly greater.

Finally, for the execution time, it is important to consider that comparison between our experiments and those made by other studies is only possible if the same software and hardware requirements are met. Otherwise, only the difference in execution time between algorithms of the same study can be analyzed.

The fastest algorithm is the Single-Objective GA, due to the lightweight methods and few generations that it had executed. The NSGA-II obtained the worst values, because of the additional steps executed at each iteration, and despite implementing a fast-sorting method. When compared against our GPPR proposal, the difference is significant, being the NSGA-II almost ten times slower for the smallest dataset (P1), and fairly slower for larger datasets. It is interesting to highlight that, as dataset size grows, the difference between the NSGA-II and our GPPR proposal decreases. However, MONRP instances are not expected to have a scale large enough that the NSGA-II could outperform our GPPR. Therefore, these values demonstrate that our proposal can be applied satisfactorily to MONRP reducing drastically computational cost.

Table 1. Average metrics of the best configurations for each dataset

Dataset	Algorithm	HV	Δ-Spread	Spacing	Exec. time (s)
P1	Single-objective GA	0.594↓	**0.615**	0.323↓	**17.967**
	NSGA-II	**1.0**	0.963↓	**0.382**	180.991↓
	GPPR	0.909↓	0.644	0.371↓	18.102
P2	Single-objective GA	0.157↓	**0.637**	0.128↓	**82.713**
	NSGA-II	0.407↓	0.969↓	0.245↓	616.415↓
	GPPR	**0.973**	0.688↓	**0.300**	250.841↓
S3	Single-objective GA	0.102↓	0.720	0.105↓	**125.702**
	NSGA-II	0.286↓	0.970↓	0.206↓	859.928↓
	GPPR	**0.977**	**0.711**	**0.293**	488.045↓

6   Conclusions and Future Work

In this paper, we have studied the applicability of a greedy procedure (GPPR) into a multi-objective problem of the Software Engineering field. The MONRP has been tackled previously using, mainly, evolutionary approaches. Few proposals have used GRASP-based methods, usually applying basic instances of it. Our proposal aimed to design a method capable of generating solutions of similar quality than those of the evolutive approaches, but reducing drastically the computational cost. We have explored different combinations of construction and local search methods, and applied post-construction techniques, such as PR, to improve the solutions found. To evaluate our proposal and compare it against classic methods, we have designed an experimentation framework, in which we have used two real-world datasets and created a new one synthetically, setting a rigorous experiment. The comparison have been carried out using a set of quality metrics and comparing the Pareto fronts, obtaining quite good results and showing that our GPPR proposal can outperform more classical and popular methods, in both performance and Pareto front results. Moreover, the code of the algorithms and experiments has been published to be shared by the scientific community.

In future lines of work, we will explore other approaches to the MONRP. It would also be interesting to try combining our GPPR with a post-optimization phase using an evolutive algorithm capable of enhance former solutions. Additionally, it could be interesting to implement interactions between requirements, which is of interest when projects use long-term planning.

References

1. Bagnall, A.J., Rayward-Smith, V.J., Whittley, I.M.: The next release problem. Inf. Softw. Technol. **43**(14), 883–890 (2001)
2. Baker, P., Harman, M., Steinhöfel, K., Skaliotis, A.: Search based approaches to component selection and prioritization for the next release problem. In: 2006 22nd IEEE International Conference on Software Maintenance, pp. 176–185 (2006)
3. Chaves-González, J.M., Pérez-Toledano, M.A., Navasa, A.: Software requirement optimization using a multiobjective swarm intelligence evolutionary algorithm. Knowl.-Based Syst. **83**(1), 105–115 (2015)
4. Deb, K., Pratap, A., Agarwal, S., Meyarivan, T.: A fast and elitist multiobjective genetic algorithm: NSGA-II. IEEE Trans. Evol. Comput. **6**(2), 182–197 (2002)
5. del Sagrado, J., del Águila, I.M., Orellana, F.J.: Ant colony optimization for the next release problem: a comparative study. In: 2010 Proceedings - 2nd International Symposium on Search Based Software Engineering SSBSE, pp. 67–76 (2010)
6. Durillo, J.J., Zhang, Y., Alba, E., Nebro, A.J.: A study of the multi-objective next release problem. In: 2009 Proceedings - 1st International Symposium on Search Based Software Engineering SSBSE (2009)
7. Feo, T., Resende, M.: Greedy randomized adaptive search procedures. J. Global Optim. **6**, 109–133 (1995). https://doi.org/10.1007/BF01096763
8. Finkelstein, A., Harman, M., Mansouri, S.A., Ren, J., Zhang, Y.: A search based approach to fairness analysis in requirement assignments to aid negotiation, mediation and decision making. Requirements Eng. **14**(4), 231–245 (2009)
9. Greer, D., Ruhe, G.: Software release planning: an evolutionary and iterative approach. Inf. Softw. Technol. **46**, 243–253 (2004)
10. Laguna, M., Marti, R.: GRASP and path relinking for 2-layer straight line crossing minimization. INFORMS J. Comput. **11**(1), 44–52 (1999)
11. del Sagrado, J., del Águila, I., Orellana, F.J.: Multi-objective ant colony optimization for requirements selection. Empir. Softw. Eng. **20**, 577–610 (2015). https://doi.org/10.1007/s10664-013-9287-3
12. Schott, J.: Fault tolerant design using single and multicriteria genetic algorithm optimization. Ph.D. thesis, Massachusetts Institute of Technology, M.S., USA (1995)
13. Zhang, Y., Harman, M., Mansouri, A.: The multi-objective next release problem. In: GECCO 2007: Genetic and Evolutionary Computation Conference, pp. 1129–1137 (2007)
14. Zitzler, E., Thiele, L.: Multiobjective evolutionary algorithms: a comparative case study and the strength Pareto approach. IEEE Trans. Evol. Comput. **3**(4), 257–271 (1999)

Comparing Parallel Surrogate-Based and Surrogate-Free Multi-objective Optimization of COVID-19 Vaccines Allocation

Guillaume Briffoteaux[1,3]([✉]), Romain Ragonnet[2], Pierre Tomenko[1],
Mohand Mezmaz[1], Nouredine Melab[3], and Daniel Tuyttens[1]

[1] Mathematics and Operational Research Department, University of Mons,
Mons, Belgium
{guillaume.briffoteaux,pierre.tomenko,mohand.mezmaz,
daniel.tuyttens}@umons.ac.be
[2] School of Public Health and Preventive Medicine, Monash University,
Melbourne, Australia
romain.ragonnet@monash.edu
[3] University of Lille, Inria, , UMR 9189 - CRIStAL, Lille, France
nouredine.melab@univ-lille.fr

Abstract. The simulation-based and computationally expensive problem tackled in this paper addresses COVID-19 vaccines allocation in Malaysia. The multi-objective formulation considers simultaneously the total number of deaths, peak hospital occupancy and relaxation of mobility restrictions. Evolutionary algorithms have proven their capability to handle multi-to-many objectives but require a high number of computationally expensive simulations. The available techniques to raise the challenge rely on the joint use of surrogate-assisted optimization and parallel computing to deal with computational expensiveness. On the one hand, the simulation software is imitated by a cheap-to-evaluate surrogate model. On the other hand, multiple candidates are simultaneously assessed *via* multiple processing cores. In this study, we compare the performance of recently proposed surrogate-free and surrogate-based parallel multi-objective algorithms through the application to the COVID-19 vaccine distribution problem.

1 Introduction

In this paper, we address a multi-objective (MO) COVID-19 vaccines allocation problem. We aim to identify vaccines allocation strategies that minimize the total number of deaths and peak hospital occupancy, while maximizing the extent to which mobility restrictions can be relaxed. The onset of the COVID-19 outbreak has been rapidly followed by the development of dedicated simulation software to predict the trajectory of the disease [1,2]. The availability of such tools enables one to inform authorities by formulating and solving optimization

B. Dorronsoro et al. (Eds.): OLA 2022, CCIS 1684, pp. 201–212, 2022.
https://doi.org/10.1007/978-3-031-22039-5_16

problems. In [3], a SEIR-model (Susceptible, Exposed, Infectious, Recovered) is deployed to simulate COVID-19 impacts. A single-objective (SO) problem is subsequently derived and handled by grid-search to regulate the alleviation of social restrictions. Multiple SO optimizations are carried out independently by a simplex or a line search algorithm in [4–6] to efficiently allocate doses of vaccines to the age-categories of a population. The number of infections, deaths and hospital admissions are considered as the possible objective. The prioritization rules approved by the government of the studied cohort are integrated as constraints in the linear programming model presented in [7] to minimize mortality. In [8], multiple indicators are combined into a scalar-valued objective function. The MO formulation exhibited in [9] consists in maximizing the geographical diversity and social fairness of the distribution plan. Nevertheless, the MO problem is scalarized into a SO one that is then solved by a simplex algorithm. The approach used by Bubar and colleagues [10] is significantly different to ours, as the authors predefined a set of vaccination strategies and selected the most promising approach among them. In contrast, our continuous optimisation approach automatically designs strategies in a fully flexible way. The optimisation problem solved by McBryde and colleagues [11] is closer to that presented in our work since a similar level of flexibility was allowed to design optimal vaccines allocation plans. However, the authors used a simpler COVID-19 model resulting in significantly shorter simulation times, such that optimisation could be performed using more classical techniques. To the best of our knowledge, it has not been suggested yet to simultaneously minimize the number of deaths, peak hospital occupancy and the degree of mobility restriction through a MO-formulated problem. The fact that we consider the level of restrictions as one of the objectives to minimise represents a novelty compared to the previous works.

Despite the relative computational expensiveness of infectious disease transmission simulators, surrogate-based optimization has been rarely applied to the field. In [12], we harnessed surrogate models to determine the allocation of preventive treatments that minimize the number of deaths caused by tuberculosis in the Philippines. The identification of the regime for tuberculosis antibiotic treatments with lowest time and doses is formulated as a SO problem in [13] and solved by a method relying on a Radial Basis Functions surrogate model. The work presented in [14] deviates from this present study in that it aims to conceive a model prescribing the actions to perform according to a given situation. It is actually more related to artificial neural network hyper-parameters and architecture search. What is called "surrogate" in [14] is actually denominated "simulator" in simulation-based optimization. In this work, we combine machine learning and parallel computing to solve the MO vaccine distribution problem.

This study demonstrates the suitability of parallel surrogate-based multi-objective optimization algorithms on the real-world problem of COVID-19 vaccines allocation. The COVID-19-related problem is detailed in Sect. 2 and the MO algorithms are exposed in Sect. 3. Both surrogate-based and surrogate-free parallel MO approaches are applied to the real-world challenge in Sect. 4 and empirical comparisons are realized. Finally, conclusions are drawn in Sect. 5 and suggestions for future investigations are outlined.

2   COVID-19 Vaccine Distribution Problem

The vast vaccination programs implemented over the last year or so all around the world achieved dramatic reductions of COVID-19 hospitalizations and deaths [15]. However, access to vaccination remains challenging, especially for low- to middle-income countries that are not able to offer vaccination to all their citizens [16]. The problem we are concerned with consists in optimizing the age-specific vaccines allocation plan to limit the impact of the disease in Malaysia under a capped number of doses. The population is divided into 8 age-categories of 10-years band from 0–9 years old to 70+ years old and the impact is expressed in terms of total number of deaths and peak hospital occupancy.

The simulation is realized in three phases by the AuTuMN software publicly available in https://github.com/monash-emu/AuTuMN/. The simulator is calibrated during the first phase with data accumulated from the beginning of the epidemic to the 1st of April 2021. The second phase starts at this latter date and lasts three months during which a daily limited number of doses is shared out among the population. Relaxation of mobility restrictions marks the kick-off of the third phase in the course of which a new distribution plan is applied involving the same number of daily available doses as in phase 2.

Decision variables $x_i \in [0,1]$ for $1 \leqslant i \leqslant 8$ and for $9 \leqslant i \leqslant 16$ represent the proportions of the available doses allocated to the 8 age-categories for phase 2 and phase 3 respectively. Variable $x_{17} \in [0,1]$ expresses the degree of relaxation of mobility restrictions where $x_{17} = 0$ leaves the restrictions unchanged and $x_{17} = 1$ means a return back to the pre-covid *era*. The following convex constraints convey the limitation of the number of doses during phases 2 and 3:

$$\sum_{i=1}^{8} x_i \leqslant 1 \text{ and } \sum_{i=9}^{16} x_i \leqslant 1 \tag{1}$$

The three-objective optimization problem consists in finding x^* such that

$$x^* = \underset{x \in [0,1]^{17} \text{ s.t. } (1)}{\arg\min} (g_1(x), g_2(x), 1 - x_{17}) \tag{2}$$

where $g_1(x)$ is the simulated total number of deaths and $g_2(x)$ the simulated maximum number of occupied hospital beds during the period.

3   Parallel Multi-objective Evolutionary Algorithms

3.1   Variation Operators of Evolutionary Algorithms

Evolutionary Algorithms (EAs) are harnessed to deal with the COVID-19-related problem exhibited previously. In EAs, a population of solutions is evolved through cycles of parents selection, reproduction, children evaluation and replacement. EAs are chosen because they have proven their effectiveness on numerous multi-objective real-world problems [17] where the objective functions are black-box as it is the case in our scenario. The constraint being convex and analytically verifiable, it is thus possible to design specific reproduction operators that directly generate feasible candidates. Assuming that every feasible solution can be reached, this technique has shown to be a reliable one [18].

The specific cross-over operator, called *distrib-X*, considers the two phases and the degree of relaxation independently. For two parents x and y, the last decision variable for the two children z and t is set such that $z_{17} = x_{17}$ and $t_{17} = y_{17}$. Regarding the second phase, let I and J be a random partition of $\{1, \ldots, 8\}$. For the age categories in I, z receives the proportion of vaccines from x ($z_i = x_i$ for $i \in I$). The remaining proportion of available doses at this step is $r = 1 - \sum_{i \in I} x_i$. For the age categories in J, the remaining proportion of doses is shared out according to the proportion allocated to the corresponding age categories in y. In other terms, for $j \in J$, $z_j = \frac{r \cdot y_j}{\sum_{j \in J} y_j}$. A similar treatment is applied to the variables associated to the third phase. The second child t is generated with an analogous procedure, where the roles of the parents are reversed.

The specific mutation operator, denoted *distrib-M*, disturbs a decision variable randomly chosen with uniform probability for $\{1, \ldots, 8\}$, $\{9, \ldots, 16\}$ and $\{17\}$. The last decision variable is mutated by polynomial mutation [17]. For the remaining ones, two age categories of the same phase are randomly selected and a random amount of doses are transferred from the first category to the second one. Both *distrib-X* and *distrib-M* are inspired by [12].

The *intermediate* and the *2-points* cross-over operators [17] are also considered for the sake of comparison. The *intermediate* strategy combines parents by random weighting average, while the *2-points* operator distributes portions of parents to the children. The portions are defined by two points with the first one separating phase 2 and phase 3 and the second one located between phase 3 and the relaxation decision variable x_{17}.

3.2   Parallel Multi-objective Evolutionary Algorithms

The major challenge in multi-objective optimization is to balance convergence and diversity in the objective space. Convergence is related to the closeness to the Pareto Front (PF) [17]. The PF is the set of the overall best solutions represented in the objective space and the Non-Dominated Fronts (NDFs) are approximations of the PF. Diversity is indicated by an extended coverage of the objective space by the NDFs. Hereafter, we present four algorithms to set this trade-off.

The first algorithm considered in the comparison is the Non-dominated Sorting Genetic Algorithm (NSGA-II) [19]. Firstly, to promote convergence, solutions pertaining to better NDFs are better ranked. Secondly, to favor diversity, solutions composing the same NDF are distinguished by setting the promise as high as the crowding distance is high. The proposed sorting is employed at the selection and the replacement steps of the EA.

The second algorithm reproduced for the experiments is the Reference Vector guided Evolutionary Algorithm (RVEA) proposed in [20] to handle many-objective problems. A set of reference vectors is introduced in order to decompose the objective space and to enhance diversity. New candidates are attached to their closest reference vector, thus forming sub-populations among which only one candidate is kept at the replacement step. The new angle penalized distance chooses adaptively the candidate to be conserved by favoring convergence at the beginning of the search and diversity at latter stages. It is worth noting that the population size may change during the search in RVEA due to the possibility of empty sub-populations. In cases of degenerated or disconnected PF a high number of sub-populations become empty and the NDF obtained at the end of the search may not be dense enough. In the RVEA* variant, an additional reference vectors set is used to replace the reference vectors that would correspond to empty sub-populations.

In surrogate-based optimization, the additional trade-off between exploitation and exploration is to be specified. Minimizing the predicted objective vectors (POVs) produced by the surrogate boosts exploitation of known promising regions of the search space. Conversely, maximizing the predictive uncertainty enhances exploration of unknown regions.

The third algorithm is the surrogate-based Adaptive Bayesian Multi-Objective Evolutionary Algorithm (AB-MOEA) [21]. The first step of a cycle in AB-MOEA consists in generating new candidates by minimizing the POVs thanks to RVEA. During the second step, the new candidates are re-evaluated by an adaptive function that favors convergence at the beginning and reinforces exploitation as the execution progresses by minimizing the predictive uncertainty delivered by the surrogate. At the third step, q candidates are retained based on an adaptive sampling criterion similar to the reference vector guided replacement of RVEA to promote diversity.

The fourth algorithm is the Surrogate-Assisted Evolutionary Algorithm for Medium Scale Expensive problems (SAEA-ME) [22]. In SAEA-ME, NSGA-II is used to optimize a six-objective acquisition function where the three first objectives are the POVs and the last three objectives are the POVs minus the predictive variances. From the set of proposed candidates, the q ones showing the best hyper-volume improvement considering both the POVs alone and the POVs minus two variances are retained for parallel simulations. SAEA-ME performs well on problems with more than 10 decision variables. The dimensionality reduction feature proposed in [22] is not considered here as it consumes computational budget and can be applied to any method.

A Multi-Task Gaussian Process (MTGP) surrogate model [23] is implemented *via* the GPyTorch library [24] and incorporated into both AB-MOEA and SAEA-ME. Using a MTGP to model multiple objectives has been realized in [25] to control quality in sheet metal forming. In a traditional regression GP [26], a kernel function is specified to model the covariance between the inputs, thus allowing the model to learn the input-output mapping and to return predictions and predictive uncertainties. In the MTGP, inter-task dependencies are also taken into account in the hope of improving over the case where the tasks are decoupled.

In the present investigation, the tasks are the three objectives and five kernel functions are considered for comparison. The widely used Radial Basis Functions kernel, denoted *rbf* and described in [26], provides very smooth predictors. According to [27], the Matern kernel with hyper-parameter $\nu = 1.5$ or 2.5, called *matern1.5* and *2.5* respectively, is to be preferred to model many physical phenomena. The higher predictive capacity Spectral Mixture kernel proposed in [28] is also raised with 2 and 4 components, denominated *sm2* and *sm4* respectively.

4   Experiments

The computational budget is set to two hours on 18 computing cores, thus allowing 18 simulations to be realized in parallel. The simulation duration varies from one solution to another from 13 to 142 s on one computing core. The four competing algorithms are implemented using our pySBO Python tool publicly available at: https://github.com/GuillaumeBriffoteaux/pySBO. Ten repetitions of the searches are carried out to ensure statistical robustness of the comparisons. The reference point for hyper-volume calculation is set to an upper bound for each objective $(32.10^6; 32.10^6; 1.5)$.

The surrogate-free approaches NSGA-II, RVEA and RVEA* are equipped with either the *distrib-X*, the *2-points* or the *intermediate* cross-over operator. For NSGA-II, the population size p_s is set to 108 or 162, thus avoiding the idling of the computing cores. For RVEA and RVEA*, we choose $p_s = 105$ or 171 to comply with the constraint imposed by the reference vectors initialization and to keep values close to those imposed for NSGA-II. Ten initial populations composed of 171 simulated solutions are generated to start the algorithms. Each initial population is made at 85% of solutions randomly sampled within the feasible search space and at 15% of candidates picked out on the boundary. When $p_s < 171$, only the best p_s candidates according to the non-dominated sorting defined in [19] are retained. For RVEA and RVEA*, a scaled version of the problem, where the first two objectives are divided by 1000, is also considered to demonstrate the effect of the objectives scales on the behavior of the methods.

The surrogate-based approaches AB-MOEA and SAEA-ME only integrate the *distrib-X* operator and use all the 171 initial samples as initial database. For RVEA in AB-MOEA, $p_s = 105$ and the number of generations is fixed to 20 as recommended in [21], while $p_s = 76$ for NSGA-II in SAEA-ME according to the guidance provided in [22] and the population evolved for 100 generations.

Table 1 shows the ranking of the algorithms according to the final hyper-volumes averaged over the ten repetitions. It can be observed in Table 1 that all the surrogate-based strategies outperform all those without surrogate. In particular, SAEA-ME with the *matern1.5* kernel is the best approach. The MTGP equipped with the *matern1.5* covariance function is preferred in both the SAEA-ME and AB-MOEA frameworks. Regarding the surrogate-free methods, NSGA-II with the *distrib-X* cross-over mechanism and $p_s = 108$ yields the best averaged hyper-volume. It is worth noticing that the *distrib-X* operator, specifically designed for the problem at hand, is to put forward as it surpasses both the *intermediate* and the *2-points* strategies in all contexts. Among the RVEAs, the best variant is RVEA* with $p_s = 105$ and the *distrib-X* cross-over thus indicating a possibly degenerated or disconnected PF. Indeed, the PF is certainly degenerated as indicates Fig. 1 where are plotted the objective vectors from the ten final NDFs obtained by SAEA-ME with the *matern1.5* kernel. When analyzing the influence of objectives scales over the efficiency of RVEA and RVEA*, the conclusions drawn in [20] are confirmed as both algorithms are more appropriate when objectives have similar scales. Indeed, the three objectives lie in $[1655; 13, 762]$, $[843; 10, 962]$ and $[0; 1]$, respectively. The previous ranges are approximated *a posteriori* based on 250,664 simulations performed in RVEA and RVEA* on the original problem. The necessity to adequately scale the objectives brings a disadvantage to RVEAs as the scaling weights are tedious to define especially in the context of black-box expensive simulations. Another drawback is the constraints on the population size preventing to totally impede the idling of computing cores in all scenarios.

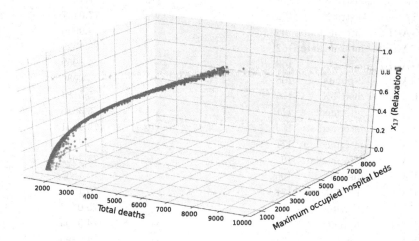

Fig. 1. Best NDFs from the 10 repetitions for SAEA-ME with matern1.5 kernel.

Figure 2 monitors the averaged hyper-volume as the search proceeds for the best strategy per category according to Table 1. The hyper-volume improves sharply at the very beginning of the search for the surrogate-based methods and reaches convergence rapidly (around 300 to 500 simulations). NSGA-II converges much slower but seems not to have converged at the end of the execution. By the right extremities of the curves, it could be expected that the hyper-volume returned by NSGA-II exceeds the one from AB-MOEA for larger numbers of simulations. However, reiterating the experiments for a time budget of four hours has

Table 1. Ranking of the surrogate-based and surrogate-free approaches according to the averaged final hyper-volumes over the 10 repetitions.

Algorithm	Cross-over operator	Population size	GP kernel	Objectives scaling	Averaged final Hyper-volume $(\times 10^{10} + 1.535 \times 10^{15})$
SAEA-ME	**distribX**	**76**	**matern1.5**	–	**80.1800**
SAEA-ME	distribX	76	matern2.5	–	80.1610
SAEA-ME	distribX	76	rbf	–	79.9541
SAEA-ME	distribX	76	sm2	–	79.6701
AB-MOEA	distribX	105	matern1.5	–	79.6200
AB-MOEA	distribX	105	matern2.5	–	79.5879
SAEA-ME	distribX	76	sm4	–	79.5789
AB-MOEA	distribX	105	sm4	–	79.4861
AB-MOEA	distribX	105	sm2	–	79.4841
AB-MOEA	distribX	105	rbf	–	79.4304
NSGA-II	distribX	108	–	–	79.3337
NSGA-II	distribX	162	–	–	79.1876
RVEA*	distribX	105	–	yes	77.2805
RVEA*	distribX	171	–	yes	77.2514
RVEA	distribX	171	–	yes	77.1287
RVEA	distribX	105	–	yes	77.0117
NSGA-II	intermediate	108	–	–	76.9946
NSGA-II	intermediate	162	–	–	76.8320
NSGA-II	2-points	162	–	–	75.6959
NSGA-II	2-points	108	–	–	75.5889
RVEA	distribX	105	–	–	75.5184
RVEA*	intermediate	171	–	yes	75.3816
RVEA*	intermediate	105	–	yes	75.2841
RVEA	intermediate	171	–	yes	75.2006
RVEA	intermediate	105	–	yes	75.1562
RVEA*	distribX	105	–	–	75.1555
RVEA*	distribX	171	–	–	75.1372
RVEA	distribX	171	–	–	75.0563
RVEA*	2-points	171	–	yes	74.9803
RVEA	2-points	171	–	yes	74.9195
RVEA	2-points	105	–	yes	74.7692
RVEA*	2-points	105	–	yes	74.7535
RVEA*	intermediate	105	–	–	74.5607
RVEA	intermediate	105	–	–	74.5585
RVEA	intermediate	171	–	–	74.4959
RVEA*	intermediate	171	–	–	74.4266
RVEA	2-points	171	–	–	74.3694
RVEA	2-points	105	–	–	74.3518
RVEA*	2-points	171	–	–	74.3264
RVEA*	2-points	105	–	–	74.2507

not allowed to verify this expectation. Figure 2 specifies that the impact of objectives scaling on RVEAs appears from around 300 simulations. In the setting of a capped computational budget, it is important to strongly favor convergence and exploitation at the onset of the search. SAEA-ME and AB-MOEA realizes this by minimizing the POVs at the top beginning of the execution. The difference between the two approaches lies in the incorporation of the predictive uncertainty. In SAEA-ME, a degree of exploration is maintained by maximization of the predictive variance. Conversely, minimization of the predictive uncertainty is involved at latter stages in AB-MOEA. In spite of the convergence-oriented strategy adopted by RVEAs at the early stages of the search, the embedded mechanism set up to handle many objectives is quite heavy and reveals to be unsuitable when the computational budget is restricted. Indeed, in [20] the algorithms are run from 500 to 1,000 generations while 10 to 20 generations are allowed by our computational budget.

Reducing the solving time of moderately expensive optimization problems where the simulation lasts less than five minutes may enable to manage optimization under uncertainty. As the calibration of the simulation tool is uncertain, multiple configurations of its parameters can be considered, resulting in multiple optimization exercises to be executed and thus enabling to gain insight about the variability of the results.

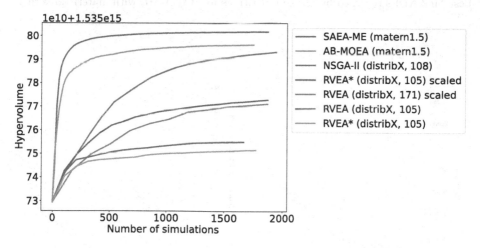

Fig. 2. Averaged hyper-volume according to the number of simulations.

The optimal allocation plan implies providing 70% of the doses to the 10–19 years old age-group and 30% to the 20–29 age-group during phase 2 according to Fig. 3. In phase 3, 70% of the doses are assigned to 20–29 years old individuals and 15% to both the 40–49 and 10–19 age-categories. This plan prioritizes the vaccination of younger adults as they are the most transmitting cohort because of their high contact rate in the population [29]. Nevertheless, the present results have to be considered with caution. Since our experiments date back to the beginning of 2021, few feedback about vaccination efficiency was available. It is assumed here that the vaccine reduces transmission although it might not be the case for the Omicron variant of concern that started to break through the world at the end of 2021. Our results are similar to those presented in [30,31] for influenza. From Fig. 4 where the total number of deaths and the maximum number of occupied hospital beds are displayed with respect to the relaxation variable x_{17}, the alleviation of the physical distancing reveals to trigger an augmentation of the hospital occupancy and deaths.

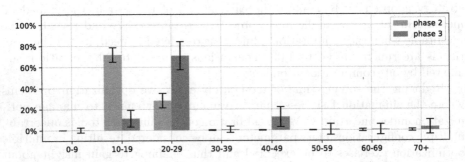

Fig. 3. Vaccines distribution according to age-categories. Averaged solutions from the best final NDFs returned by the 10 repetitions for SAEA-ME with matern1.5 kernel.

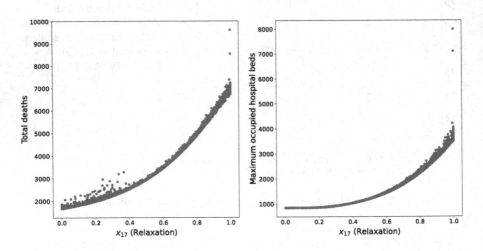

Fig. 4. Total number of deaths and maximum number of occupied hospital beds according to relaxation of the physical distancing x_{17}. Best NDFs from the 10 repetitions for SAEA-ME with matern1.5 kernel.

5   Conclusion

This paper demonstrates the suitability of parallel surrogate-based multi-objective optimization algorithms to handle the moderately computationally expensive COVID-19 vaccines allocation problem for Malaysia. In particular, SAEA-ME provides reliable results in a fast way. As future works, we suggest to benefit from the computational cost reduction of black-box simulation-based problem solving to take the uncertainty around the calibration of the simulator into account.

References

1. Chang, S., et al.: Modelling transmission and control of the COVID-19 pandemic in australia. Nat. Commun. 11,5710, 03 (2020)
2. Trauer, J.M., et al.: Understanding how Victoria, Australia gained control of its second COVID-19 wave. Nat. Commun. **12**(6266), 1–10 (2021)
3. Duque, D., Morton, D.P., Singh, B., Du, Z., Pasco, R., Meyers, L.A.: Timing social distancing to avert unmanageable COVID-19 hospital surges. Proc. Natl. Acad. Sci. **117**(33), 19873–19878 (2020)
4. Matrajt, L.: Optimizing vaccine allocation for COVID-19 vaccines: potential role of single-dose vaccination. Nat. Commun. 12(3449) (2021)
5. Matrajt, L., Longini, I.: Optimizing vaccine allocation at different points in time during an epidemic. PloS one, **5**(11), e13767 2010
6. Matrajt, L., Eaton, J., Leung, T., Brown, E.R.: Vaccine optimization for COVID-19: who to vaccinate first? Sci. Adv. **7**(6), eabf1374 (2021)
7. Buhat, C., et al.: Using constrained optimization for the allocation of COVID-19 vaccines in the Philippines. Appl. Health Econ. Health Policy **19**(5), 699–708 (2021)
8. Han, S., et al.: Time-varying optimization of COVID-19 vaccine prioritization in the context of limited vaccination capacity. Nat. Commun. **12**(1), 4673 (2021)
9. Anahideh, H., Kang, L., Nezami, N.: Fair and diverse allocation of scarce resources. Socio-Econ. Plann. Sci. **80**, 101193 (2021)
10. Bubar, K.M., et al.: Model-informed COVID-19 vaccine prioritization strategies by age and serostatus. Science **371**(6532), 916–921 (2021)
11. McBryde, E.S., et al.: Modelling direct and herd protection effects of vaccination against the SARS-CoV-2 delta variant in Australia. Med. J. Aust. **215**(9), 427–432 (2021)
12. Briffoteaux, G., et al.: Parallel surrogate-assisted optimization: batched Bayesian neural network-assisted GA versus q-ego. Swarm Evol. Comput. **57**, 100717 (2020)
13. Cicchese, J.M., Pienaar, E., Kirschner, D.E., Linderman, J.J.: Applying optimization algorithms to tuberculosis antibiotic treatment regimens. Cell. Mol. Bioeng. **10**(6), 523–535 (2017)
14. Miikkulainen, R., et al.: From prediction to prescription: evolutionary optimization of nonpharmaceutical interventions in the COVID-19 pandemic. IEEE Trans. Evol. Comput. **25**(2), 386–401 (2021)
15. Vilches, T.N., et al.: COVID-19 hospitalizations and deaths averted under an accelerated vaccination program in northeastern and southern regions of the USA. Lancet Reg. Health - Am. **6**, 100147 (2022)

16. Sheel, M., McEwen, S., Davies, S.E.: Brand inequity in access to COVID-19 vaccines. Lancet Reg. Health - W. Pac. **18**, 100366 (2022)
17. Talbi, E.G.: Metaheuristics: From Design to Implementation. Wiley, Wiley Series on Parallel and Distributed Computing (2009)
18. Michalewicz, Z., Dasgupta, D., Le Riche, R.G., Schoenauer, M.: Evolutionary algorithms for constrained engineering problems. Comput. Ind. Eng. **30**(4), 851–870 (1996)
19. Deb, K., Pratap, A., Agarwal, S., Meyarivan, T.: A fast and elitist multiobjective genetic algorithm: NSGA-ii. IEEE Trans. Evol. Comput. **6**(2), 182–197 (2002)
20. Cheng, R., Jin, Y., Olhofer, M., Sendhoff, B.: A reference vector guided evolutionary algorithm for many-objective optimization. IEEE Trans. Evol. Comput. **20**(5), 773–791 (2016)
21. Wang, X., Jin, Y., Schmitt, S., Olhofer, M.: An adaptive Bayesian approach to surrogate-assisted evolutionary multi-objective optimization. Inf. Sci. **519**, 317–331 (2020)
22. Ruan, X., Li, K., Derbel, B., Liefooghe, A.: Surrogate assisted evolutionary algorithm for medium scale multi-objective optimisation problems. In: Proceedings of the 2020 Genetic and Evolutionary Computation Conference. GECCO 2020, pp. 560–568, New York, NY, USA, Association for Computing Machinery (2020)
23. Bonilla, E.V., Chai, K., Williams, C.: Multi-task gaussian process prediction. In: Advances in Neural Information Processing Systems, vol. 20. Curran Associates Inc (2008)
24. Gardner, J.R., Pleiss, G., Bindel, D., Weinberger, K.Q., Wilson, A. G.: Gpytorch: blackbox matrix-matrix gaussian process inference with GPU acceleration. In: Advances in Neural Information Processing Systems (2018)
25. Xia, W., Yang, H., Liao, X., Zeng, J.: A multi-objective optimization method based on gaussian process simultaneous modeling for quality control in sheet metal forming. Int. J. Adv. Manufact. Technol. **72**, 1333–1346 (2014)
26. Rasmussen, C.E.: Gaussian Processes for Machine Learning. MIT Press, Cambridge (2006)
27. Stein, M.L.: Interpolation of Spatial Data: Some Theory for Kriging. Springer, New York (1999)
28. Wilson, A.G., Adams, R.P.: Gaussian process kernels for pattern discovery and extrapolation (2013)
29. Prem, K., Cook, A.R., Jit, M.: Projecting social contact matrices in 152 countries using contact surveys and demographic data. PLoS Comput. Biol. **13**(9), 1–21 (2017)
30. Weycker, D., et al.: Population-wide benefits of routine vaccination of children against influenza. Vaccine **23**(10), 1284–1293 (2005)
31. Medlock, J., Galvani, A.P.: Optimizing influenza vaccine distribution. Science **325**(5948), 1705–1708 (2009)

Decentralizing and Optimizing Nation-Wide Employee Allocation While Simultaneously Maximizing Employee Satisfaction

Aniqua Tabassum[1,2]($\boxtimes$) (ID), Ashna Nawar Ahmed[2] (ID), Subhe Tasnia Alam[2] (ID),
Noushin Tabassum[2] (ID), and Md. Shahriar Mahbub[2] (ID)

[1] Seattle University, Seattle, USA
atabassum@seattleu.edu
[2] Ahsanullah University of Science and Technology, Dhaka, Bangladesh
ashna.cse@aust.edu, subhe321tasmia@gmail.com,
noushintabassum89@gmail.com, shahriar.cse@aust.edu

Abstract. Despite the studies on human resource allocation, a problem of dissatisfied employees arises in developing and under-developed countries while decentralizing human resources nationwide since rural areas have fewer facilities than urban areas. Randomly allocating employees contributes to employees' dissatisfaction if they are displeased with where they are assigned, leading to an unstable work environment. However, allocating employees solely based on their satisfaction may lead to a centralized solution around the urban cities. Therefore, employee satisfaction and dispersion are the two most essential but opposing factors for employee decentralization in developing countries.

In this study, we have addressed the problem of employee decentralization by proposing a Multi-Objective Optimization approach that maximizes the two conflicting objectives: employee satisfaction (ES) and employee dispersion (ED). A neural network is applied that predicts the ES of an employee allocated in an area/city. Moreover, we have formulated a dispersion function that provides a score based on how well dispersed a specific allocation is. Using a Multi-Objective evolutionary algorithm, we have developed an allocation framework that maximizes these conflicting objectives and finds optimal allocations.

Keywords: Employee dispersion · Employee satisfaction · Human resource allocation in developing countries · Multi-objective evolutionary algorithm

1 Introduction

Non-optimal worker allocation in developing countries is a crucial problem to be solved. There are studies on identifying the problems regarding resource allocation and on finding the optimal solutions [1,2]. However, these studies do not

B. Dorronsoro et al. (Eds.): OLA 2022, CCIS 1684, pp. 213–225, 2022.
https://doi.org/10.1007/978-3-031-22039-5_17

address the problem of declining employee satisfaction and productivity [3] when the workforce is decentralized throughout developing and underdeveloped countries. In developing countries, such as Bangladesh [4], rural areas are primarily under-developed as compared to urban areas, motivating the majority of people to move to cities for their livelihood [5]. Consequently, the workforce tends to be centralized around cities [6]. These problems are more observed explicitly among essential workers such as doctors [7], who are often forced to work in rural areas [8]. Most people designated to such areas show acute urgency to move away posthaste [9] owing to extreme dissatisfaction, causing a higher turnover rate and an unstable work environment [10]. On the other hand, the critical problem with allocation solely based on employee satisfaction is that the majority of the professionals would be more interested in working in the more facilitated areas [5]. Thus, the density of worker allocation would be skewed in favor of regions with more amenities, which will result in a vacuum of professionals in underdeveloped regions. Therefore, we need to implement a method to ensure the proper balance between worker dispersion and satisfaction.

To tackle the problem, we first predict employee satisfaction allocated to a particular area/city using Multi-Layer Perceptron (MLP) with nine influencing factors. In addition to this, we formulated a dispersion function that calculates the decentralization score of an allocation[1]. These two objectives are contradictory, especially for developing countries such as Bangladesh; hence, we use Multi-Objective Optimization to simultaneously maximize these objectives.

2   State of the Art

There have been multiple studies on resource allocation, and human resource allocation [1,2]. The author of [1] has proposed a multi-objective optimization solution aiming at minimizing total cost resulting from resource overallocation, project deadline exceedance, and day-by-day resource fluctuations. In the study [2], the author has shown how the Multi-Objective Particle Swarm Optimization Algorithm can be used for multi-criteria human resource allocation. Another study [11] focuses on minimizing total cost and the total time of logistic relief operation in emergencies.

Several studies can be found that focus on worker and resource distribution in countries such as Nigeria [12] and China [13]. The author of [12] focused on the factors that hinder recruitment and retention of the healthcare workforce, such as insufficient infrastructure, inadequately trained staff, sub-optimal distribution of healthcare workers, mainly in the rural areas, and suggested approaches to improve this situation in Nigeria. The authors of [13] pointed out the ethical flaws in various Government policies leading to the inequality in resource distribution and have proposed countermeasures that optimize resource allocation, such as formulating better policies, strengthening the responsibilities of both governmental and public financial investments, improving the utilization of resources.

---

[1] An allocation refers to the assignments of m number of employees to n number of cities.

Two works are found that focused on the satisfaction of customers [14] and employees [15]. The author of [14] proposed a Satisfaction Function (SF) which is based on the customers' attitude regarding the products the company is offering. The authors of [15] proposed to make the work environment suitable enough such that the employees can have more decision-making power which will increase sustainable practices and improve employee satisfaction. Another study [16] focused on finding the relationship between employee engagement and job satisfaction. They have also extracted seven job satisfaction factors, such as work culture and fairness at work. In [17], employee satisfaction has been predicted by a machine learning model by analyzing employees' reviews.

To best our knowledge, although there have been several works on human resource allocation, the problem of maximizing human resource dispersion (i.e., decentralization) while also maximizing employee satisfaction in their newly designated area has not been addressed yet. There have been studies regarding identifying job-satisfaction influencing factors inside organizations [17], but none of these studies focuses on identifying the influencing factors of employee satisfaction when they are allocated to new geographical areas.

3   Methodology

To maximize the two conflicting objectives of the problem: employee satisfaction and dispersion, we formulate the problem as a multi-objective optimization problem. We develop a learning model for our satisfaction predictor and formulate the dispersion function. Finally, optimized solutions are found by using a multi-objective evolutionary algorithm. The details of our methodology are presented in this chapter.

3.1   Problem Formulation

Please consider that we want to allocate m number of people to n number of areas. We have to find optimal solutions that maximize the total satisfaction of m people while fulfilling all n cities' demands of employees as much as possible. In order to solve this, we use the following two inputs in our optimization framework:

- **Cities or areas where we want to allocate our employees:** The number of vacancy each of these cities have for a given profession.
- **The people we want to allocate to different cities:** Individual description of each of these people as required by the SF (more description in 3.4).

In order to represent the first input, we considered an array where the indexes represent the cities and the values in the indexes represent the maximum capacity or requirement/demand of the corresponding city has for a specific profession. We call this our **"Capacity Array"**. We demonstrate a figure representing our Capacity Array in Fig. 1, where $a_0 - a_5$ denotes the considered six areas. For example, area number 0, or a_0 has six vacancies, a_1 has seven vacancies, and so on. Similarly, our second input is represented by an array that we call "Employee Array", where each value represents the information of an individual employee.

Index (City)	0	1	2	3	4	5
Value (Capacity)	6	7	3	5	1	2

Fig. 1. Capacity array

3.2   Problem Representation

We represent the problem as an array of size m, where the value of the specific element of the array can be 1 to n. For example, a value of 5 at the index 1 of the array means employee number 5 is assigned to city number 1. An illustration is shown below in Fig. 2.

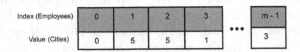

Fig. 2. Output of optimization framework

3.3   Overview of the Objectives

In this section, we have briefly described both of our objectives: employee satisfaction and dispersion.

- **Satisfaction:** The first objective is the maximization of employee satisfaction. Satisfaction of one employee refers to a number denoting how satisfied the employee is, on a scale of 1 to 5, when he/she is assigned to a certain area. If we consider that there are m employees, and if we denote individual satisfaction as s_m then for an allocation, the total satisfaction, S is

$$S = \sum_{i=1}^{m} s_m \tag{1}$$

- **Dispersion:** Our second objective is to increase the dispersion of a solution, which refers to allocating as well-spread as possible. A trend in developing countries [18] employees are satisfied only when they are designated to developed regions. Therefore, maximizing satisfaction reduces dispersion.

3.4   Modelling Satisfaction Function

In this section, we will discuss how features for our machine learning model is selected for the satisfaction prediction of an employee. A social experiment is conducted for feature selection and the data collection process. Afterward, machine learning algorithms are used for modeling employee satisfaction.

Feature Selection. In order to design the SF, we first needed to explore the factors/features that influence satisfaction. To ensure an unbiased process, we performed a social experiment by randomly choosing 20 unrelated people. We asked them one question: which factors would affect their living satisfaction in an area outside the capital. We completed the experiment without the respondents knowing which parts of their answers we would use in our research to avoid any bias. We present the results below.

- No one would move to any city with high crime rates.
- Seventeen of them would be happier if closer to their families.
- Three of the five married people did not want to move without their spouses.
- Four out of the five married people additionally mentioned that having good schools in their area would be vital to them.

Upon introducing the topic of average house rent of the city, as some cities have higher house rent, all of them agreed to this being a significant issue to consider. Their answers were varied based on gender, age, and occupation. Therefore, nine factors/features are chosen from this social experiment to use in our data collection process: gender, age, occupation, security, house rent, travel time from hometown, schooling, marital status, and spouses' willingness to relocate with their partners.

Creating Questionnaires. There being no curated dataset to train our satisfaction predictor, we created our own dataset by circulating questionnaires. We have categorized gender, age, occupation, marital status, and spouses' willingness to relocate with their partners as demographic factors. The rest four factors are scenario-based.

These remaining four scenario-based factors have three levels: low, medium, and high. The respondents have been presented with a scenario by combining the different levels of the four features and asked to provide a satisfaction score.

There are a total of 3^4 or eighty-one possible scenarios. It was impractical to ask to score each respondent all 81 virtual scenarios. Therefore, we had to carefully partition the larger set of eighty-one scenarios into smaller groups, each consisting of three unique scenarios. Thus, $(81/3) = 27$ sets of questionnaires were formed, with each one consisting of identical demographic questions and three unique scenario-based questions. The three scenarios in each set were noticeably different enough, so the respondents could easily differentiate them. A sample set of three virtual scenarios representing three areas is as follows:

- low security, medium schooling facilities, low house rent, medium travel time from hometown
- medium security, low schooling facilities, medium house rent, low travel time from hometown
- high security, high schooling facilities, low house rent, high travel time from hometown

We asked the participants to rate their satisfaction on a scale of 1 to 5 for each scenario.

It was vital to evenly distribute these 27 sets among the respondents as we want to collect respondents' satisfaction for all data points. Therefore, we created a website where we uniformly sampled the scenarios presented to users to ensure balanced distribution. We have collected 855 data points in total.

Necessity of Learning Models/Algorithms. It is necessary to predict employee satisfaction for our allocation framework. SF modeling problem can be considered as a regression problem. However, satisfaction cannot be represented as a linear combination of its features because it is a complicated and non-linear psychological issue that can vastly vary even within the same demographic. These correlations can be identified by advanced machine learning models. Therefore, We have experimented with various machine learning regression algorithms (presented in Sect. 4.1) to see which one performs best for our case.

3.5 Modelling Dispersion Function

The task of the dispersion function (DF) is to measure the distribution of workers in different areas/cities in a numeric value. If there are four cities and workers are assigned in two cities while there is no allocation in the other two cities, the dispersion value will be lower than the scenario where workers are assigned in all four cities. We also formulated the function so that it has a relation with the number of required personnel in each area. We have denoted the dispersion of an occupation p by $D(p)$, and the function can be represented as follows:

$$D(p) = |R^{(p)}| + \sum_{i=1}^{n} d_i^p \tag{2}$$

Here,

$$R^{(p)} = \{\, i \,|\, 0 \leq i \,\&\, al_i^p \geq minreq_i^p \,\} \tag{3}$$

$$minreq_i^p = \left\lfloor \frac{t^p * \frac{c_i^p}{\sum_{j=1}^{n} c_j^p} * 100}{100} \right\rfloor \tag{4}$$

$$= \left\lfloor t_p * \frac{c_i^p}{\sum_{j=1}^{n} c_j^p} \right\rfloor \tag{5}$$

$$d_i^p = \begin{cases} \frac{al_i^p}{c_i^p}, & \text{if} \quad al_i^p \leq c_i^p \\ 1 - \frac{(al_i^p - c_i^p)}{c_i^p}, & \text{otherwise} \end{cases} \tag{6}$$

where,
n = Carnality of the set of areas brought into consideration
$R^{(p)}$ = Set of regions/areas where minimum requirement of allocation of profession p have been met
d_i^p = dispersion of an area i of profession p

$minreq_i^p$ = Minimum number of employees of profession p that needs to be allocated in area i

al_i^p = number of person allocated of profession p in area i

c_i^p = Capacity or requirements of number of profession p in area i

t^p = total number of available employees of profession p

The dispersion of an area d_i is the ratio between the number of allocated persons and the requirements. If the number of allocations met requirements, we would achieve area-wise maximum distribution, which is 1. However, the algorithm may allocate more people than the areas' requirements. Then, we subtract a penalty from the maximum dispersion.

Moreover, in most developing countries, there is a shortage of skilled employees. Therefore, the total number of required people ($\sum_{j=1}^{n} c_j^p$) is higher than the available number of people (t_p). Therefore, we added a secondary mechanism (i.e., $minreq$) to calculate the minimum possible people the algorithm can allocate in an area given that there are t_p number of professionals.

Finally, when calculating total dispersion score, the following two factors are added:

- The number of cities/areas where the minimum number of required employees for that city/area has been allocated
- Summation of all the area-wise dispersion

The theoretical maximum dispersion value can be achieved if for each city, its exact number of $minreq$ professionals are assigned to it.

Sample Example. In Table 1, we have presented seven cities indexed from zero to six. We have mentioned their required number of employees, and have calculated their minimum requirements, d_i^p and if their demands have been met.

Table 1. Calculating minimum requirement, d_i^p and R^p

City no	Capacity	Allocation	minreq	d_i^p	R^p
0	6	4	4	0.66	0
1	10	7	7	0.7	1
2	8	6	6	0.75	2
3	10	7	7	0.7	3
4	3	3	2	1	4
5	4	3	3	0.75	5
6	8	6	3	0.75	6

Therefore, $R^{(p)} = \{0, 1, 2, 3, 4, 5, 6\}$ and $|R^{(p)}| = 7$. Thus, dispersion, $D(p) = |R^{(p)}| + \sum_{i=1}^{n} d_i^p = 7 + 5.316 = 12.316$

4   Experiments and Results

For implementing the SF, we used scikit-learn [19], a python-based machine learning framework. Implementation[2] of multi-objective optimization for finding optimum allocation using SF and DF has been done using jMetalPy [20], a python-based multi-objective meta-heuristic framework. The results of the prediction of SF of different learning algorithms are presented. Different hyperparameter settings are explored to find the best possible settings. Finally, the results of two optimization algorithms are shown.

4.1   Results of Satisfaction Modelling

This section shows the results of different learning models. We also present some of the hyperparameter settings we have experimented with to find the best setting for the model. In Table 2 we have presented the different methods and their $R2$ scores [21] on training and cross validation (CV) dataset.

Table 2. R2 score of training and cross validation (CV) for different methods

Name	Train	CV
Linear regression [22]	40.01%	42.1%
Linear regression in log-space [23]	42.6%	45.3%
Random forest [24]	74.8%	34.0%
2-degree polynomial regression [25]	47.5%	48.1%
3-degree polynomial regression [26]	57.8%	0.43%
Multi layer perceptron regression [27]	**51.3%**	**49.7%**

We can see that MLP performs the best R2 score in both training and cross-validation datasets combined. So, MLP Regression is the best-suited method for this problem as it has the highest R2 score. Getting an R2 score higher than 0.5 is hard when the model includes human psychology because each human thinks differently, so it is typically hard to generalize it [28].

Afterward, we do hyperparameter tuning by performing several experiments with different parameter settings for the MLP model. The dataset was standardized and normalized before performing the experiments and running the iterations until convergence. We present the results in Table 3.

---

[2] https://github.com/aniquaTabassum/Undergrad-Thesis.

Table 3. R2 score of training and cross validation

Batch size	Optimizer	Activation	Layers	Neurons per layer	LR	Train	CV
8	**LBFGS**	**ReLU**	**3**	**(100, 8, 8)**	**0.0013**	**51.30%**	**49.74%**
8	LBFGS	tanh	3	(100, 8, 8)	0.0013	45.06%	46.83%
8	Adam	ReLU	3	(100, 8, 8)	0.0013	−80.0%	−35.0%
8	LBFGS	ReLU	3	(100, 8, 8)	0.0003	50.95%	48.89%
16	LBFGS	ReLU	5	(100, 16, 16, 8, 8)	0.0013	49.53%	49.53%
4	LBFGS	ReLU	4	(8, 8, 8, 25)	0.0013	49.9%	45.0%

As per Table 3, we chose a 3-layer MLP regressor with ReLU activation function in the hidden layer, LBFGS optimizer, and a learning rate of 0.0013 as our model, as this performs the best on both training and cross-validation datasets.

4.2   Results of Optimization

This section presents a scenario of a doctor allocation problem in Bangladesh. We chose the medical profession for our optimization experiments since they are categorized as one of the foremost essential workers [7] and are needed in every region of a country. However, our framework works appropriately for other professions as well. Afterward, optimization results of the problem are presented.

Scenario: A Doctor Allocation Problem. Thirty-six doctors filled out our survey, and we have applied our framework to allocate all of them in the seven major cities of Bangladesh. Even with just 36 employees being allocated to 7 cities, there are 7^{36} possible allocation combinations, which is impossible to simulate using an exhaustive search. We collect the necessary information (i.e., security, house rent, travel time from hometown, schooling) for the seven cities, and the features of the cities are labeled by different markers (i.e., low, medium, and high) based on the collected information. For measuring schooling facilities, the results of the Secondary School Certificate (SSC) exam [29] of the schools in each of the cities for the year 2019 [30] are considered. We took the median house rent [31] for each of these cities. Security is estimated by the total crime rate for each of these cities in 2019 calculated from the data [32]. The maximum time (data taken from Google Map [33]) required to reach one city to another is taken as the travel time between two cities. We prefer travel time over road distance (in KM) because the road distance does not reflect the real situation in Bangladesh as road traffic condition is very poor[3].

Finally, the features of the cities are ranked in low, medium, and high. A feature value fall within bottom 33.33% is "low", a value in between 33.33% and 66.66% is "medium", and a value over 66.66% is considered as "high".

---

[3] Road distance will perform similarly if the road traffic condition is similar to all over the region.

Results. We compare the performance of the two most used multi-objective optimization algorithms (NSGA-II [34], and SPEA2 [35]) on our problem. Two algorithms separately ran fifty times to make a statistically significant comparison. The parameter settings for the experiment are given below in Table 4.

Table 4. Parameter setting for optimal algorithm

Parameter	NSGA-II	SPEA-II
Population size	100	100
Crossover type	Single-point	Single-point
Crossover probability	0.9	0.9
Mutation type	Random mutation	Random mutation
Mutation probability	0.027[a]	0.027
Max evaluations	30000	30000

[a] $mutation probability = \frac{1}{number\ of\ decision\ variables} = \frac{1}{36} = 0.027$

Figure 3 shows the approximated true Pareto-front of the problem. Since the true Pareto-front is unknown, we ran the algorithm 50 times, merged the results, and created an approximation of the true Pareto front.

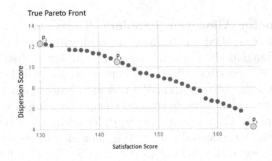

Fig. 3. Approximation of true Pareto-front

In Fig. 3, point p_1 is an extreme solution with the highest satisfaction value, 166 out of 185, but has the lowest dispersion value, 4.25 out of the theoretical maximum value of 12.31. This solution can be chosen when ES is far more important than ED. Point p_3 has the lowest satisfaction value, 130 out of 166, and the highest possible dispersion score, 12.31. Such a solution is ideal if achieving maximum dispersion is the goal. Point p_2 is a random solution in the Pareto-front. Here, the satisfaction score is 143, and the dispersion score is 10.61.

Boxplots of hypervolume [36,37], IGD [36,38] and spread [36,39] are presented in Fig. 4 where we can see that the two algorithms are producing similar results. It can be concluded that either algorithm could be chosen.

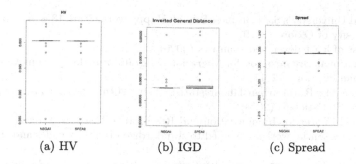

(a) HV   (b) IGD   (c) Spread

Fig. 4. Comparison between NSGA-II and SPEA-II

5 Conclusion

In this study, we have proposed a more practical approach for human resource allocation for developing and under-developed countries, maximizing employee satisfaction and dispersion, essential for stabilizing the workforce problem in rural areas. First, we have identified the satisfaction influencing factors for employees and have gathered a dataset accordingly to create our satisfaction predictor. Then we mathematically formulated employee dispersion to ensure fair distribution of employees. Our proposed dispersion function also considers the shortage of human resources. A multi-objective evolutionary algorithm approach is applied to find the optimal set of solutions, among which any solution can be chosen based on the situation at hand. On our dataset, the highest satisfaction score achieved by our framework is 166 out of 185 (i.e., 36 doctors can have maximum satisfaction of 5), and the average satisfaction score of 36 employees is 4.61 out of 5.0, which is very high. Even in a solution where the dispersion score is 10.61 out of 12.31 (i.e., maximum dispersion can be calculated theoretically, please see dispersion modeling Sect. 3.5), the average satisfaction of 36 employees is as high as 3.9. We plan to improve our satisfaction model by feeding it more data in the future. The proposed allocation framework is generalized and can be applied for building an organized and content set of workforce, since decentralizing employees is becoming a necessity for companies as they keep expanding, be that among the cities of a country or worldwide.

References

1. Kaiafa, S., Chassiakos, A.P.: A genetic algorithm for optimal resource-driven project scheduling. Procedia Eng. **123**, 260–267 (2015). Selected papers from Creative Construction Conference 2015
2. Jia, Z., Gong, L.: Multi-criteria human resource allocation for optimization problems using multi-objective particle swarm optimization algorithm. In: 2008 International Conference on Computer Science and Software Engineering, vol. 1, pp. 1187–1190. IEEE (2008)

3. Oxford University's Saïd Business School: Happy workers are 13% more productive. University of Oxford (2019)
4. UN list of least developed countries (2020)
5. International Organization for Migration: Migration: making the move from rural to urban by choice (2017)
6. Castillo, M.: Rural-urban labour statistics. In: 20th International Conference of Labour Statisticians (2018)
7. National Center for Immunization and Respiratory Diseases: Interim list of categories of essential workers mapped to standardized industry codes and titles. CDC (2021)
8. Dasgupta, K.: Young doctors don't want to go to rural areas because they feel ill-equipped (2016)
9. Darkwa, E.K., Newman, M., Kawkab, M., Chowdhury, M.E.: A qualitative study of factors influencing retention of doctors and nurses at rural healthcare facilities in Bangladesh. BMC Health Serv. Res. **15**(1), 1–12 (2015)
10. Nanda, A., Soelton, M., Luiza, S., Saratian, E.T.P.: The effect of psychological work environment and work loads on turnover interest, work stress as an intervening variable. In: 4th International Conference on Management, Economics and Business (ICMEB 2019), pp. 225–231. Atlantis Press (2020)
11. Sarma, D., Bera, U.K., Das, A.: A mathematical model for resource allocation in emergency situations with the co-operation of NGOs under uncertainty. Comput. Ind. Eng. **137**, 106000 (2019)
12. Awofeso, N.: Improving health workforce recruitment and retention in rural and remote regions of Nigeria. Rural Remote Health **10**(1), 162–171 (2010)
13. Chen, Y., Yin, Z., Xie, Q.: Suggestions to ameliorate the inequity in urban/rural allocation of healthcare resources in China. Int. J. Equity Health **13**, 34 (2014)
14. Pereira, A.C.: A mathematical model to measure customer satisfaction. Qual. Eng. **11**(2), 281–286 (1998)
15. Vinerean, S., Cetina, I., Dumitrescu, L.: Modeling employee satisfaction in relation to CSR practices and attraction and retention of top talent. Expert J. Bus. Manage. **1**(1), 4–14 (2013)
16. Singh, L.B.: Job satisfaction as a predictor of employee engagement. Amity Glob. Bus. Rev. **7**(1), 20–30 (2017)
17. Conlon, S., Simmons, L., Liu, F.: Predicting tech employee job satisfaction using machine learning techniques. Int. J. Manage. Inf. Technol. **16**, 72–88 (2021)
18. Manzi, L.: Migration from rural areas to cities: challenges and opportunities (2020)
19. Scikit-Learn Contributors: Scikit-learn machine learning in Python (2021)
20. jMetalPy Contributors: jMetalPy (2021)
21. Minitab Blog Editor: Regression analysis: how do i interpret r-squared and assess the goodness-of-fit? (2013)
22. Yale University: Linear regression (1997)
23. Benoit, K.: Linear regression models with logarithmic transformations. London Sch. Econ. **22**(1), 23–36 (2011)
24. Yiu, T.: Understanding random forest - towards data science (2021)
25. Abhigyan: Understanding polynomial regression!!! - Analytics Vidhya (2020)
26. StatsDirect: Polynomial regression
27. Wikipedia Contributors: Multilayer perceptron (2021)
28. Callum Ballard: An ODE to R-squared (2021)
29. Wikipedia Contributors: Secondary school certificate (2020)
30. SSC Result 2019 (2020)

31. Bproperty: Location based house rent search (2020)
32. Crime statistics 2019 (2020)
33. Google Maps, Bangladesh (2020)
34. Deb, K., Pratap, A., Agarwal, S., Meyarivan, T.: A fast and elitist multiobjective genetic algorithm: NSGA-II. IEEE Trans. Evol. Comput. **6**(2), 182–197 (2002)
35. Zitzler, E., Laumanns, M., Thiele, L.: SPEA2: improving the strength Pareto evolutionary algorithm. TIK-report **103** (2001). Eidgenössische Technische Hochschule Zürich (ETH), Institut für Technische Informatik und Kommunikationsnetze (TIK)
36. Audet, C., Bigeon, J., Cartier, D., Le Digabel, S., Salomon, L.: Performance indicators in multiobjective optimization. Eur. J. Oper. Res. **292**(2), 397–422 (2021)
37. Auger, A., Bader, J., Brockhoff, D., Zitzler, E.: Hypervolume-based multiobjective optimization: theoretical foundations and practical implications. Theoret. Comput. Sci. **425**, 75–103 (2012)
38. Sun, Y., Yen, G.G., Yi, Z.: IGD indicator-based evolutionary algorithm for many-objective optimization problems. IEEE Trans. Evol. Comput. **23**(2), 173–187 (2019)
39. Li, M., Zheng, J.: Spread assessment for evolutionary multi-objective optimization. In: Ehrgott, M., Fonseca, C.M., Gandibleux, X., Hao, J.-K., Sevaux, M. (eds.) EMO 2009. LNCS, vol. 5467, pp. 216–230. Springer, Heidelberg (2009). https://doi.org/10.1007/978-3-642-01020-0_20

Categorical-Continuous Bayesian Optimization Applied to Chemical Reactions

Theo Rabut[1(✉)], Hamamache Kheddouci[1], and Thomas Galeandro-Diamant[2,3]

[1] Université de Lyon, Université Lyon 1, LIRIS UMR CNRS 5205,
69621 Lyon, France
`theo.rabut@univ-lyon1.fr`
[2] ChemIntelligence, Lyon, France
[3] DeepMatter, Glasgow, UK

Abstract. Chemical reaction optimization is a challenging task for the industry. Its purpose is to experimentally find reaction parameters (e.g. temperature, concentration, pressure) that maximize or minimize a set of objectives (e.g. yield or selectivity of the chemical reaction). These experiments are often expensive and long (up to several days), making the use of modern optimization methods more and more attractive for chemistry scientists.

Recently, Bayesian optimization has been shown to outperform human decision-making for the optimization of chemical reactions [16]. It is well-suited for chemical reaction optimization problems, for which the evaluation is expensive and noisy.

In this paper we address the problem of chemical reaction optimization with continuous and categorical variables.

We propose a Bayesian optimization method that uses a covariance function specifically designed for categorical and continuous variables and initially proposed by Ru *et al.* in the COCABO method [14].

We also experimentally compare different methods to optimize the acquisition function. We measure their performances in the optimization of multiple chemical reaction (or formulation) simulators.

We find that a brute-force approach for the optimization of the acquisition function offers the best results but is too slow when there are many categorical variables or categories. However we show that an ant colony optimization technique for the optimization of the acquisition function is a well-suited alternative when the brute-force approach cannot be (reasonably) used.

We show that the proposed Bayesian optimization algorithm finds optimal reaction parameters in fewer experiments than state of the art algorithms on our simulators.

Keywords: Mixed Bayesian optimization · Chemical reaction optimization · Categorical variables

© The Author(s), under exclusive license to Springer Nature Switzerland AG 2022
B. Dorronsoro et al. (Eds.): OLA 2022, CCIS 1684, pp. 226–239, 2022.
https://doi.org/10.1007/978-3-031-22039-5_18

1   Introduction

Every chemical reaction is optimized before being industrialized. The goal is to find, by carrying out experiments, input parameters (e.g. temperature, pressure, residence time, etc.) that offer optimal values for a set of objectives (e.g. maximize the yield, minimize the production of an impurity, etc.).

The pursuit of high-performance optimization methods is driven by the high cost of chemical experiments. The performances of optimization methods applied to chemical reactions are measured against the quality of the solution (i.e. how close the solution is to the optimization objectives) and how many experiments are needed to find this solution.

One-Variable-At-a-Time (OVAT) and Design of Experiments (DoE) [1,18] methods are the most used approaches to optimize chemical reactions. The OVAT method iterates by performing experiments and modifying only one parameter at a time. DoE methods consist in planning a series of experiments following a design matrix, running these experiments and building a statistical model (usually linear or polynomial) with the resulting dataset. An optimum is then computed from the model. OVAT and DoE methods tend to need a large number of experiments to be effective. In addition, OVAT can be very slow (because only one variable is changed at a time) and can get stuck in local optima. Simplex-based methods are also sometimes used to optimize chemical reactions [11,21]. They consists of building a simplex in the search space, then evaluating the objective function at each of the vertices of the simplex and iteratively displacing one vertex at a time following heuristics. Simplex-based methods tend to be easily stuck in local optima [20].

Zhou et al. [23] proposed a deep reinforcement learning (DRL) based method to optimize chemical reactions. The authors combined DRL and pre-training to be able to start working with very small amounts of data. This leads to satisfactory results on problems containing only continuous variables but hasn't been tested with categorical variables (without descriptors).

Bayesian optimization (BO) is a powerful approach to optimize problems for which the evaluations are expensive and noisy. It has shown a variety of successful applications [15]. BO concepts are described in Fig. 1. First, an initialisation is done with a small number of experiments. Then, a surrogate model (e.g. Gaussian process) is trained using these experiments. An acquisition function, that balances the predicted improvement (exploitation strategy) and the uncertainty of the predictions (exploration strategy), is applied to the model. An optimization algorithm is applied to find the maximum of this acquisition function. The set of parameters that gives this maximal value for the acquisition function determines the next experiment (chemical reaction) to run. This experiment is run, its result is added to the dataset, and the algorithm starts a new iteration. The algorithm stops when the objectives are attained or when the experiments budget is spent.

Categorical variables are often present in the optimization of chemical reactions [13]. We can cite as an example the choice of a catalyst or additives, the choice of the solvent or the order of addition of the reactants. Categorical

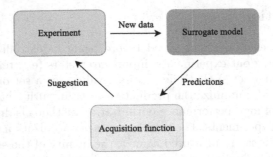

Fig. 1. Simplified Bayesian optimization algorithm applied to chemical reactions.

variables have two important particularities. The first one is the non-continuity constraint, since categorical variables are not defined on a continuous space. The second one is the non-ordinality constraint: they can only be compared with the equality operator. For example, with a categorical variable representing a choice between the solvents *water, ethanol, toluene*, asserting that *water > toluene* is meaningless.

Mixed-variable optimization can be handled with one-hot encoding: a categorical variable with n categories is encoded as a vector of n corresponding bits, with all bits being equal to 0 except the bit corresponding to the selected category, that is equal to 1. However, in the BO algorithm, treating one-hot dimensions as continuous without any supplementary treatment misleads the acquisition function optimizer and often results in a sub-optimal solution. Indeed, the experiment proposed by the acquisition function optimizer is a real-valued vector and has to be decoded to the closest category. Hence, most of the time, there will be a gap between the experiment suggested by the acquisition function optimizer and the experiment that will actually be performed, leading to a mediocre optimization performance.

The work presented by Garrido-Merchán *et al.* [4] brings an improvement to the basic one-hot encoding approach. During the optimization of the acquisition function, real-valued encoded vectors are transformed to the nearest one-hot vectors before being used as inputs of the model. It follows that the acquisition function optimizer considers real-valued vectors as having the same acquisition values as the associated transformed vectors. Thus, the acquisition optimizer suggests an experiment that can be performed as is, which ensures the convergence to optimal solutions.

Häse *et al.* [5] have developed an augmented Bayesian optimization algorithm called Gryffin that uses a Bayesian neural network as surrogate model. It estimates kernel densities, based on previously evaluated experiments, that are used to approximate the objective function. Gryffin is able to use expert knowledge (descriptors) to guide the optimization, which drastically improves the performances of their method. Its "naive" version doesn't use descriptors, which enabled us to use it in our benchmarks.

COCABO [14] is a Bayesian optimization method designed for mixed-variable optimization. At each iteration, COCABO first selects categories with a multi-armed bandit algorithm and then separately optimizes the numerical variables (after modelling them using a mixed covariance function).

Random forests can be used as surrogate model in Bayesian optimization instead of Gaussian processes [6]. A ready-to-use implementation of this app-roach is provided in a package called SMAC [8] which handles categorical vari-ables.

In this study, we aim at improving the performances (i.e. reducing the num-ber of experiments necessary to reach an optimum) of the Bayesian optimization method for the optimization of chemical reactions with continuous and categor-ical variables. Our approach is based on Gaussian processes as surrogate models with the COCABO covariance function [14]. We propose different techniques for the optimization of the acquisition function. Next, we compare the different acquisition function optimizer on the optimization of simulated chemical reac-tions. And finally, we compare our optimization algorithm (using the COCABO covariance function and the highest-performing acquisition function optimizer) with other state-of-the-art algorithms.

2   Problem Definition

Our work is applied to problems with a form given by:

$$\text{Minimize } f(\mathbf{z}) \text{ with the smallest possible number of evaluations} \tag{1}$$

where:

- $\mathbf{z} = (\mathbf{x}, \mathbf{h})$
- $\mathbf{x} = x_0, .., x_n$ and $x_i \in [A_i, B_i]$ with $A_i, B_i \in \mathbb{R}$
- $\mathbf{h} = h_0, ..., h_n$ and $h_i \in C_i$ with C_i denotes the categorical space of the ith categorical variable.

This work is restricted to single objective optimization. Moreover, only contin-uous and categorical variables are used.

The "No-Free Lunch Theorem" [19] stipulates that the performances of every optimization methods are equal when averaged on all possible problems. It implies that in order to increase the performances on a specific optimization problem (e.g. chemical reaction optimization), we must evaluate the optimiza-tion method on similar problems without any regards on the performances of unrelated ones. The underlying functions of chemical reactions have some par-ticularities: they are smooth and have few local optima [10,17]. So, in order to be specific to the chemical reaction optimization problem, we measure the performances of our approach using chemical reaction (or formulation) simu-lators. We have built these simulations by training machine learning models with publicly available chemical data (see Table 1). Each data set has been har-vested from patents or academic articles and are produced by the optimiza-tion of chemical reactions (or chemical formulations). The individuals of these data sets corresponds to experiments where input parameters (e.g. temperature, choice of catalyst) have been tested and scores have been calculated following the observations made at the end of these experiments. In the case of the first Suzuki-Miyaura reaction simulation (second row of the Table 1), there are two objectives (turnover number and yield). We choose to optimize the mean of the two scores.

This benchmarking strategy was initially introduced by Felton et al. [3] for measuring performances on chemical reactions with continuous and categorical variables. It allows us to establish optimization performances on chemical reactions without having to run experiments in a chemistry lab.

Table 1. Details of the data used to train the simulators

Reaction type	Number of experiments	Source
Pd-catalysed direct arylation	1728	[16]
Suzuki-Miyaura cross-coupling	4 cases of 96	[3,13]
Stereoselective Suzuki-Miyaura cross-coupling	192	[2]
Polycarbonate resin formulation	100	[9]

3   Propositions

In a first part, we describe the surrogate model including the COCABO kernel and its hyperparameters. In a second part, we present different approaches for the optimization of the acquisition function.

3.1   Gaussian Process Kernel

We use Gaussian processes (GP) to approximate the underlying functions of chemical reactions. It is the most commonly used model since it can inherently predict both a value and an associated uncertainty. Gaussian processes are mainly defined by their covariance function. Since the underlying functions of chemical reactions are smooth, we use a smooth covariance function, Matérn$_{5/2}$ [12], for the continuous dimensions.

The smoothness of the GP on continuous variables is kept with the use of the one-hot encoding. However, the Euclidean distance used for the calculation of the Matérn$_{5/2}$ kernel is based on all dimensions (continuous and encoded). We believe that, in order to capture complex relationships between categorical and continuous variables, the covariance function should use the Euclidean distance only on continuous variables and incorporate categorical knowledge later in its calculation. The COCABO method [14] uses such a covariance function (see Eq. 2). It combines two sub-functions: one for continuous variables, K_{cont}, and one for categorical variables, K_{cat}.

$$K(\mathbf{z}, \mathbf{z}') = (1 - \lambda) \times (K_{cont}(\mathbf{x}, \mathbf{x}') \times K_{cat}(\mathbf{h}, \mathbf{h}'))$$
$$+ \lambda \times (K_{cont}(\mathbf{x}, \mathbf{x}') + K_{cat}(\mathbf{h}, \mathbf{h'})) \qquad (2)$$

where:

- $\mathbf{z} = (\mathbf{x}, \mathbf{h})$
- $\mathbf{x}$ is the set of continuous variables
- $\mathbf{h}$ is the set of categorical variables

K_{cont} is the Matérn$_{5/2}$ function. It is a standard covariance function for smooth Gaussian processes regressions with continuous inputs. K_{cat}, the kernel for categorical inputs (see Eq. 3), measures similarity between categorical vectors with the equality operator (which is the only permitted operation for categorical variables).

$$K_{cat}(\mathbf{h}, \mathbf{h}') = \sigma \times \frac{1}{D} \sum_{1}^{D} \alpha(h_d, h'_d) \qquad (3)$$

where:

- $\alpha(a, b)$ equals 1 if $a = b$ and 0 if $a \neq b$
- D is the number of categorical variables
- σ is the variance hyperparameter.

The proposition made by Ru *et al.* in COCABO [14] revolves around the hyperparameter λ, which is a trade-off between the two terms of the Eq. 2: the sum and the product of K_{cont} and K_{cat}. Both of these terms capture different relationships between continuous and categorical variables. The sum of the two sub-kernels produces a learning of a single trend on the continuous variables and shift this trend depending on the categories whereas the product is able to produce a learning of complex relationships with highly different trends depending on the categories. The sum is especially necessary when the amount of training data is low (beginning of the optimization) because the product is able to capture knowledge only if the evaluations have categories in common. For example, if two evaluations have the same continuous features but different categorical ones, the product will be equal to 0 which prevent the model to learn even on continuous variables. Nonetheless, the product is essential because, as the optimization goes on, more evaluations are added to the training dataset and a single trend with a simple shift will not be sufficient to model the complexity offered by the data. In other words the sum alone will not be able to capture all the knowledge available to guide the optimization. With the hyperparameter λ, the authors ensure that the relationships that can be captured either by the sum or by the product are taken into account into the covariance $K(\mathbf{z}, \mathbf{z}')$, because λ is tuned during the fitting of the Gaussian process.

In order to avoid underfitting/overfitting the data while training the Gaussian process (tuning its hyperparameters to minimize its negative log marginal likelihood [12]), we confined hyperparameter values within a range. σ_K, $\sigma_{K_{cont}}$

and $\sigma_{K_{cat}}$ were bounded in $[10^{-2}, 20]$ while the lengthscale parameter of K_{cont} and λ were respectively bounded in $[10^{-2}, 20]$ and $[0.1, 0.9]$. We used the L-BFGS optimizer to tune the GP hyperparameters.

3.2 Acquisition Function Optimization

We chose to use the Expected Improvement (EI) acquisition function because it has shown good results on diverse applications and has a strong theoretical support [22]. The equation of Expected Improvement is given by:

$$\mathrm{EI}(\mathbf{x}) = \mathbb{E}[\max(f(\mathbf{x}) - f(\mathbf{x}^+), 0)] \tag{4}$$

with $f(\mathbf{x}^+)$ the value of the evaluation that have yielded the best result so far. The analytical form of EI is the following:

$$\mathrm{EI}(\mathbf{x}) = \begin{cases} \sigma(\mathbf{x})Z\Phi(Z) + \sigma(\mathbf{x})\phi(Z) & \text{if } \sigma(\mathbf{x}) \neq 0 \\ 0 & \text{if } \sigma(\mathbf{x}) = 0 \end{cases} \tag{5}$$

where

$$Z = \frac{\mu(\mathbf{x}) - f(\mathbf{x}^+) - \xi}{\sigma(\mathbf{x})} \tag{6}$$

$\Phi(Z)$ and $\phi(Z)$ denotes respectively the cumulative distribution function (CDF) and the probability density function (PDF) of the variable Z. Z denotes the predicted improvement divided by the standard deviation (uncertainty) and the parameter ξ determines the weight of the exploration strategy in the equation. This analytical form of EI is cheap to evaluate and can be optimized without sparing on the number of evaluations. Therefore, we propose several approaches for the optimization of the acquisition function with mixed variables.

The first approach (denoted as L-BFGS-OHE) involves the one-hot encoding of the categorical variables and a multi-started gradient descent for the optimization of the acquisition function. However, since the COCABO model does not accept one-hot vectors, one-hot dimensions are systematically decoded before any predictions. In other words, predictions are asked for by the acquisition function optimizer with encoded inputs but they are decoded before they pass through the model. The multi-started gradient descent is performed as follows: 1000 configurations are randomly drawn and the 5 configurations with the highest acquisition function value are kept and a gradient descent (L-BFGS) is performed on each of these 5 configurations.

We also propose an approach based on a "brute-force" optimization of the categorical space and a multi-started gradient descent on the continuous space (see Algorithm 1). First, all the combinations of the categorical parameters are constructed. Then, for each combination, a multi-started gradient descent (previously described) is performed on the continuous parameters. Finally, after determining the maximal acquisition values for each categorical combination, the configuration with the highest acquisition value is suggested as the next experiment. This algorithm reduces the difficulty of the optimization of the acquisition function because instead of dealing with different types of variables (or with supplementary dimensions from the encoding), the acquisition optimizer only works on the continuous dimensions. Still, it can be heavy in terms of computational cost if the number of categories and categorical variables is large.

Algorithm 1. Categorical brute-force and multi-started gradient descent

---

1: Construct all categorical combinations
2: Multi-started gradient descent optimization of continuous parameters for each combination
3: Choose as suggestion the configuration (continuous and categorical) with the highest acquisition

---

While the brute-force approach evaluates thoroughly the search space, with five optimizations of the continuous space for each categorical combination, this method becomes prohibitively long to run when the number of categorical combinations is higher than a few hundreds.

Lastly, we implemented an evolutionary algorithm based the behaviour of ant colonies (ACO) that scales better with the number of categorical variables than brute-force [7]. In our experiments, we used the colony hyperparameters proposed by the authors without any restart allowed. This algorithm is a multi-agent method inspired by the behaviour of ants. An ant represent an evaluation at a given set of parameters. At each generation, each ant randomly moves towards previously evaluated points with good results (exploitation strategy). The presence of multiple ants in the colony and the randomness of their movements enable the mandatory exploration of the search space. It allows the ants to not only moves around promising areas but also randomly explore areas that may have not been explored so far.

4   Results

This section presents the optimization of four simulators. We did 25 runs with 50 suggestions for each optimization algorithm. At the beginning of each run, we randomly drew 5 initial evaluations and, for a fairness purpose, these 5 evaluations were used to initialize all the optimizers. As shown in the Table 2, the number of input variables (16) of the polycarbonate resin formulation is higher than the other simulations (less than 6), so we allowed 100 suggestions for each optimizer on this simulation.

Table 2. Number of variables for each simulations

Reaction type	Cont. var.	Cat. var.
Pd-catalysed direct arylation	2	3
Suzuki-Miyaura cross-coupling	3	1
Stereoselective Suzuki-Miyaura cross-coupling	5	1
Polycarbonate resin formulation	11	5

4.1   Acquisition Function Optimizer

The four plots presented in Fig. 2 are used to compare the performances of different acquisition optimizers, using the COCABO kernel and the Expected Improvement acquisition function. The average (and standard deviation) of the best score obtained at the end of each optimization are summarized in Table 3. On each benchmark we provide, as a baseline, the results of a random strategy that suggests input parameters randomly.

On the first three simulators (Figs. 2a, 2b and 2c), brute-force outperforms the other methods. On the last simulator (see Fig. 2d), the brute-force approach cannot be used due to the high number of categorical variables and categories. Indeed, there are 1728 combinations of categorical variables, so the brute-force approach consists of 8640 optimization routines upon continuous variables, which is too long to run.

The approach based on categorical relaxation (L-BFGS-OHE) with the one-hot encoding and a rounding to the closest categorical variable before passing through the model, is the second best performing method on every simulators. Even if the optimizer (L-BFGS) evolves in a continuous space with a large number of encoded dimensions and a large number of flat regions (depending on the number of categorical variables and categories), the optimizer manages to optimize the acquisition function.

The behaviour of the ant colony optimization method (ACO) highly depends on its parameters, notably the exploration hyperparameter q which is set to 0.05 and the *restart* parameter. The ACO performs poorly, compared to the other optimization methods, on the simulators with a small number of dimensions but performs slightly better on the polycarbonate resin formulation simulator which have more dimensions.

As consequence of the results presented above and for the rest of our study, we choose the brute-force approach to be the acquisition function optimizer when the number of categorical variables is lower than 4. ACO is used otherwise.

4.2   Comparison with Other Methods

The next results (Fig. 3) present a comparison between our method (composed of the COCABO kernel, Expected Improvement and the brute-force optimizer or the ant colony optimizer when the brute-force method can't be used), Gryffin [5], COCABO [14], SMAC [6], and the work of Garrido-Merchán *et al.* [4].

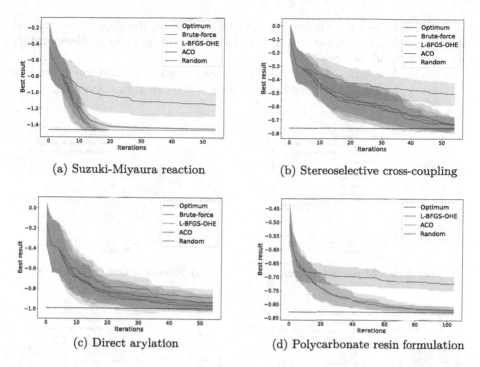

Fig. 2. Best score evolution on simulatiors with the use of different acquisition function optimizers (Brute-force, L-BFGS-OHE, ACO, Random).

Table 3. Average scores for different acquisition optimizers at 5 random initial evaluations and 50 suggestions (100 for the polycarbonate resin formulation). Best scores on each simulators are marked in bold.

Method	Suzuki-Miyaura	Stereoselective Cross coupling	Direct arylation	Polycarbonate resin formulation
Brute-force	$-\mathbf{1.460} \pm 0.00$	$-\mathbf{0.733} \pm 0.06$	$-\mathbf{0.965} \pm 0.05$	–
L-BFGS-OHE	$-\mathbf{1.460} \pm 0.00$	-0.730 ± 0.05	-0.959 ± 0.07	-0.820 ± 0.01
ACO	-1.441 ± 0.01	-0.726 ± 0.04	-0.929 ± 0.08	$-\mathbf{0.821} \pm 0.02$
Random	-1.141 ± 0.15	-0.504 ± 0.08	-0.875 ± 0.08	-0.723 ± 0.03

We used the "naive" version of Gryffin in its authors' implementation, which does not use any chemical descriptor to guide the optimization. The computation time of Gryffin is long when the number of dimensions is high, so we used the "boosted" version of Gryffin for the optimization of the polycarbonate resin formulation simulation.

We used COCABO in its authors' implementation with its default settings and a starting $\lambda = 0.5$.

SMAC denotes an optimization algorithm based on Random Forest [6] and the Expected Improvement acquisition function. We used an implementation proposed by Lindauer *et al.* [8].

"Garrido-Merchán - 2020" is a Bayesian optimization method which involves a Matérn$_{5/2}$ kernel and the one-hot encoding of the categorical variables.

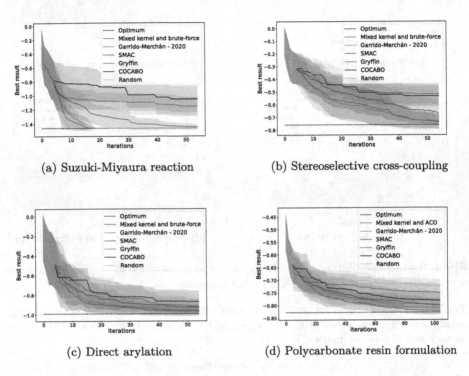

(a) Suzuki-Miyaura reaction   (b) Stereoselective cross-coupling

(c) Direct arylation   (d) Polycarbonate resin formulation

Fig. 3. Best score evolution on simulators with the use of different optimization methods: mixed kernel and brute-force (or ACO), Garrdi-Merchán - 2020, SMAC, Gryffin, COCABO, Random.

Table 4. Average scores for different optimization methods at 5 random initial evaluations and 50 suggestions (100 for the polycarbonate resin formulation). Best scores on each simulators are marked in bold.

Method	Suzuki-Miyaura	Stereoselective Cross coupling	Direct arylation	Polycarbonate resin formulation
Mixed kernel and brute-force	$-\mathbf{1.460} \pm 0.01$	-0.733 ± 0.06	$-\mathbf{0.965} \pm 0.05$	–
Mixed kernel and ACO	-1.441 ± 0.01	-0.726 ± 0.04	-0.929 ± 0.08	$-\mathbf{0.821} \pm 0.02$
Garrido-Merchán - 2020	-1.459 ± 0.01	$-\mathbf{0.738} \pm 0.05$	-0.957 ± 0.06	-0.797 ± 0.02
SMAC	-1.443 ± 0.04	-0.674 ± 0.09	-0.924 ± 0.06	-0.799 ± 0.03
Gryffin	-1.149 ± 0.13	-0.517 ± 0.08	-0.912 ± 0.07	-0.751 ± 0.03
COCABO	-1.046 ± 0.18	0.534 ± 0.08	-0.865 ± 0.11	-0.781 ± 0.03
Random	-1.160 ± 0.13	0.504 ± 0.08	-0.847 ± 0.05	-0.723 ± 0.03

The results on the simulator denoted as "Suzuki-Miyaura" (which corresponds to a simulation built upon the first data set proposed by Reizman *et al.* [13]), shows that the Mixed kernel combined with a categorical brute-force approach for the optimization of the acquisition function is the method that offers the best performances.

The method denoted as "Garrido-Merchán - 2020" is the method that provides the best performances on the second simulator ("stereoselective cross-coupling").

On the Pd-catalysed direct arylation simulator, the COCABO kernel combined with the brute-force approach gives the best results. Gryffin and COCABO do not provide satisfactory scores on the two first simulators ("Suzuki-Miyaura", "stereoselective cross-coupling"). However these optimization methods provide better results on the other two simulators.

The method composed of the COCABO kernel and the ant colony optimization technique gives the best averaged final score on the polycarbonate resin formulation simulator (-0.821).

SMAC performs poorly on the first three simulators compared to Bayesian optimization with Gaussian process as surrogate model. However on the last simulator, the random forest based Bayesian optimization performances are similar to the Bayesian optimization with one-hot encoding and the Matérn$_{5/2}$ kernel.

We show that our method composed of the COCABO kernel and either the categorical brute-force or the ant colony optimizer as the acquisition function optimization technique, generally converges faster to the optimum than the other methods.

4.3   Kernel Influence on Performances

The Table 5 summarises the averaged final scores of two Bayesian optimization methods that only differ from each other by their covariance functions and the use of the one-hot encoding. "Garrido-Merchán-2020" is using the one-hot encoding and a standard continuous kernel (Matérn$_{5/2}$) whereas the method denoted as "L-BFGS-OHE" is using a kernel specially designed for continuous and categorical variables without encoding (the COCABO kernel). Their results are similar on the first simulator but on the second one ("stereoselective cross coupling"), the method denoted as "Garrido-Merchán-2020" achieves better performances (-0.738) than "L-BFGS-OHE" method (-0.730). On the polycarbonate resin formulation simulator, the COCABO kernel based method ("L-BFGS-OHE") performs better (-0.820) than "Garrido-Merchán-2020" (-0.797). This comparison exposes that the COCABO kernel should be used instead of a Matérn$_{5/2}$ kernel and the one-hot encoding when the chemical reaction (or formulation) optimization problem involves multiple continuous and categorical variables. Indeed, this covariance function allows a more efficient Bayesian optimization on our simulators. The authors of COCABO (Ru *et al.*) attribute this phenomenon to a stronger modeling power of the COCABO kernel.

Table 5. Average final scores of "Garrido-Merchán-2020" and "L-BFGS-OHE" extracted from Table 3 and Table 4.

Method	Suzuki-Miyaura	Stereoselective Cross coupling	Direct arylation	Polycarbonate resin formulation
L-BFGS-OHE	-1.460 ± 0.00	-0.730 ± 0.05	-0.959 ± 0.07	-0.820 ± 0.01
Garrido-Merchán - 2020	-1.459 ± 0.01	-0.738 ± 0.05	-0.957 ± 0.06	-0.797 ± 0.02

5   Conclusion

This paper presents a study of different optimization techniques for the optimization of chemical reaction (or formulation) simulators with mixed variables (continuous and categorical).

We expose a Bayesian optimization algorithm based on a Gaussian process with a covariance function specifically designed for continuous and categorical variables [14]. Also, we evaluate different methods for the optimization of the acquisition function. We show that when facing a small number of categorical variables, a categorical brute-force approach associated with a multi-started gradient descent performs best for the optimization of the acquisition function. The ant colony optimization method is the method the most suited (in our study) for the optimization of the acquisition function when the number of categorical combinations is too high to use the brute-force based approach. A more in-depth study of these evolutionary methods will be the subject of further works. Our method globally performs better than other state-of-the-art methods [4,5,8,14] on our simulators.

Since the use of a kernel specifically designed for continuous and categorical inputs shows the best performances, we are working on further increasing the quality of the model by modifying the COCABO covariance function. We believe that more relationships between variables and evaluations can be captured with a new set of parameters in the covariance function.

Nevertheless, in order to fully establish the performance of the presented method, we are currently working on an experimental validation in chemistry labs.

Acknowledgement. This work was supported by the R&D Booster SMAPI project 2020 of the Auvergne-Rhône-Alpes Region.

References

1. Carlson, R., Carlson, J.E.: Design and Optimization in Organic Synthesis. Elsevier, Amsterdam (2005)
2. Christensen, M., et al.: Data-science driven autonomous process optimization. Commun. Chem. 4(1), 1–12 (2021)
3. Felton, K., Rittig, J., Lapkin, A.: Summit: benchmarking machine learning methods for reaction optimisation. Chem. Methods 1(2), 116–122 (2020)

4. Garrido-Merchán, E.C., Hernández-Lobato, D.: Dealing with categorical and integer-valued variables in Bayesian Optimization with Gaussian processes. Neurocomputing **380**, 20–35 (2020)

5. Häse, F., Aldeghi, M., Hickman, R.J., Roch, L.M., Aspuru-Guzik, A.: GRYFFIN: an algorithm for Bayesian optimization of categorical variables informed by expert knowledge. Appl. Phys. Rev. **8**(3), 031406 (2021)

6. Hutter, F., Hoos, H.H., Leyton-Brown, K.: Sequential model-based optimization for general algorithm configuration. In: Coello, C.A.C. (ed.) LION 2011. LNCS, vol. 6683, pp. 507–523. Springer, Heidelberg (2011). https://doi.org/10.1007/978-3-642-25566-3_40

7. Liao, T., Socha, K., de Oca, M.A.M., Stützle, T., Dorigo, M.: Ant colony optimization for mixed-variable optimization problems. IEEE Trans. Evol. Comput. **18**(4), 503–518 (2013)

8. Lindauer, M., et al.: SMAC3: a versatile Bayesian optimization package for hyperparameter optimization. arXiv: 2109.09831 (2021)

9. Monden, T.: Polycarbonate resin composition (2019). EP2810989B1

10. Moore, K.W., Pechen, A., Feng, X.J., Dominy, J., Beltrani, V.J., Rabitz, H.: Why is chemical synthesis and property optimization easier than expected? Phys. Chem. Chem. Phys. **13**(21), 10048–10070 (2011)

11. Morgan, S.L., Deming, S.N.: Simplex optimization of analytical chemical methods. Anal. Chem. **46**(9), 1170–1181 (1974)

12. Rasmussen, C.E., Nickisch, H.: Gaussian processes for machine learning (GPML) toolbox. J. Mach. Learn. Res. **11**, 3011–3015 (2010)

13. Reizman, B.J., Wang, Y.M., Buchwald, S.L., Jensen, K.F.: Suzuki-Miyaura cross-coupling optimization enabled by automated feedback. React. Chem. Eng. **1**(6), 658–666 (2016)

14. Ru, B., Alvi, A., Nguyen, V., Osborne, M.A., Roberts, S.: Bayesian optimisation over multiple continuous and categorical inputs. In: International Conference on Machine Learning, pp. 8276–8285. PMLR (2020)

15. Shahriari, B., Swersky, K., Wang, Z., Adams, R.P., De Freitas, N.: Taking the human out of the loop: a review of Bayesian optimization. Proc. IEEE **104**(1), 148–175 (2015)

16. Shields, B.J., et al.: Bayesian reaction optimization as a tool for chemical synthesis. Nature **590**(7844), 89–96 (2021)

17. Tibbetts, K.M., Feng, X.J., Rabitz, H.: Exploring experimental fitness landscapes for chemical synthesis and property optimization. Phys. Chem. Chem. Phys. **19**(6), 4266–4287 (2017)

18. Weissman, S.A., Anderson, N.G.: Design of experiments (DoE) and process optimization. A review of recent publications. Org. Process Res. Dev. **19**(11), 1605–1633 (2015)

19. Wolpert, D.H., Macready, W.G., et al.: No free lunch theorems for search. Tech. rep., Technical Report SFI-TR-95-02-010, Santa Fe Institute (1995)

20. Wright, M.H., et al.: Nelder, Mead, and the other simplex method. Doc. Math. **7**, 271–276 (2010)

21. Xiong, Q., Jutan, A.: Continuous optimization using a dynamic simplex method. Chem. Eng. Sci. **58**(16), 3817–3828 (2003)

22. Zhan, D., Xing, H.: Expected improvement for expensive optimization: a review. J. Global Optim. **78**(3), 507–544 (2020). https://doi.org/10.1007/s10898-020-00923-x

23. Zhou, Z., Li, X., Zare, R.N.: Optimizing chemical reactions with deep reinforcement learning. ACS Cent. Sci. **3**(12), 1337–1344 (2017)

Assessing Similarity-Based Grammar-Guided Genetic Programming Approaches for Program Synthesis

Ning Tao[1,2]([✉])[ID], Anthony Ventresque[1,2][ID], and Takfarinas Saber[1,3][ID]

[1] Lero – the Irish Software Research Centre, Limerick, Ireland
[2] School of Computer Science, University College Dublin, Dublin, Ireland
ning.tao@ucdconnect.ie, anthony.ventresque@ucd.ie
[3] School of Computer Science, National University of Ireland, Galway, Ireland
takfarinas.saber@nuigalway.ie

Abstract. Grammar-Guided Genetic Programming is widely recognised as one of the most successful approaches for program synthesis, i.e., the task of automatically discovering an executable piece of code given user intent. Grammar-Guided Genetic Programming has been shown capable of successfully evolving programs in arbitrary languages that solve several program synthesis problems based only on a set of input-output examples. Despite its success, the restriction on the evolutionary system to only leverage input/output error rate during its assessment of the programs it derives limits its scalability to larger and more complex program synthesis problems. With the growing number and size of open software repositories and generative artificial intelligence approaches, there is a sizeable and growing number of approaches for retrieving/-generating source code based on textual problem descriptions. Therefore, it is now, more than ever, time to introduce G3P to other means of user intent (particularly textual problem descriptions). In this paper, we would like to assess the potential for G3P to evolve programs based on their similarity to particular target codes of interest (obtained using some code retrieval/generative approach). We particularly assess 4 similarity measures from various fields: text processing (i.e., FuzzyWuzzy), natural language processing (i.e., Cosine Similarity based on term frequency), software clone detection (i.e., CCFinder), plagiarism detector(i.e., SIM). Through our experimental evaluation on a well-known program synthesis benchmark, we have shown that G3P successfully manages to evolve some of the desired programs with three of the used similarity measures. However, in its default configuration, G3P is not as successful with similarity measures as with the classical input/output error rate at evolving solving program synthesis problems.

Keywords: Program synthesis · Grammar-guided genetic programming · Code similarity · Textual description · Text to code

B. Dorronsoro et al. (Eds.): OLA 2022, CCIS 1684, pp. 240–252, 2022.
https://doi.org/10.1007/978-3-031-22039-5_19

1 Introduction

Genetic Programming (GP [16]) is an efficient approach to evolve code using high-level specifications, hence it is the most popular approach to tackle program synthesis problems (i.e., the task of automatically discovering an executable piece of code given user intent) for software engineering [24,25] and testing [26]. Various GP systems with different representations have been designed over time to tackle the diverse program synthesis problems.

PushGP [22] is one of the most efficient GP systems. PushGP evolves programs in the specially purpose-designed Push language. (i.e., a stack-based language designed specifically for program synthesis task). In Push, every variable type (e.g. strings, integers, etc.) has its own stack, which facilitates the genetic programming process. Despite its efficiency, PushGP's dependence to Push (a language that is not commonly used in practice and that is hard to interpret) hinders its exploitability and lowers our ability to improve upon it.

Grammar-Guided Genetic Programming (G3P [7]) system is another efficient GP system that evolves programs based on a specified grammar syntax. Besides its efficiency at solving program synthesis problems, the use of a syntax grammar enables G3P to produce programs that are syntactically correct with respect to any arbitrary programming languages definable through a grammar. The use of a grammar makes G3P particularly easy to move from one system to another and to adapt from one language to another [7]. This flexibility elevates G3P to be widely recognised as one of the most successful program synthesis approaches.

A recent comparative study [6] has evaluated the ability of both G3P and PushGP to solve several program synthesis problems from a well-studied program synthesis benchmark [10,11] based only on a set of input-output examples. The study found that G3P achieves the highest success rate at finding correct solutions when it does find any. The study also found that PushGP is able to find correct solutions for more problems than G3P, but PushGP's success rate for most of the problems was very low. However, despite G3P's and PushGP's successes, the restriction on the evolutionary systems to only leverage the input/output error rate during their assessment of the programs they derive limits their scalability to larger and more complex program synthesis problems.

Following on the big data trend [4] and the growing number and size of open software repositories (i.e., databases for sharing and commenting source code) and generative artificial intelligence approaches (generative deep learning) there is a sizeable and growing number of approaches for retrieving/generating source code based on textual problem descriptions. Therefore, it is now, more than ever, time to introduce G3P to other means of user intent (particularly textual problem descriptions). Code retrieval and code generation techniques might output several incomplete snippets or not fully fit for purpose codes–which often makes them impossible to exploit in their form. Therefore, in this work, we propose an approach whereby such code guides the search process towards programs that are similar.

In this paper, we would like to assess the potential for G3P to evolve programs based on their similarity to particular target codes of interest which would have been retrieved or generated using some particular text to code transformation. We particularly assess 4 similarity measures from various fields: text processing (i.e., FuzzyWuzzy), natural language processing (i.e., Cosine Similarity), software clone detection (i.e., CCFinder), plagiarism detector(i.e., SIM). The ultimate goal is the ability to identify the most suitable program similarity measure to guide the program synthesis search/evolutionary process when introducing code retrieval/generation from textual problem descriptions.

Through our experimental evaluation on a well-known program synthesis benchmark, we show that G3P successfully manages to evolve some of the desired programs with three of the used similarity measures. However, in its default configuration, G3P is not as successful with similarity measures as with the classical input/output error rate at evolving solving program synthesis problems. Therefore, in order to take advantage of textual problem descriptions and their subsequent text to code approaches in G3P, we need to either design better-fitted similarity measures, adapt our evolutionary operators to take advantage of program similarities, and/or combine similarity measures with the traditional input/output error rate.

The rest of the paper is structured as follows: Sect. 2 summarises the background and work related to our study. Section 3 describes our approach and details the similarity metrics used as code similarity in our evaluation. Section 4 details our experimental setup. Section 5 reports and discusses the results of our experiments. Finally, Sect. 6 concludes this work and discusses our future study.

2   Background and Related Work

In this section we present the material which forms our research background.

2.1   Genetic Programming

Genetic programming (GP) is an evolutionary approach that enables us to devise programs. GP starts with a population of random programs (often not very fit for purpose), and iteratively evolves it using operators analogous to natural genetic processes (e.g., crossover, mutation, and selection). Over the years, a variety of GP systems have been proposed–each with its specificity (e.g., GP [16], Linear GP [2], Cartesian GP [19]).

2.2   Grammar-Guided Genetic Programming

While there is a variety of GP systems, G3P is among the most successfully GP systems. What is unique to G3P is its use of a grammar as a guideline for syntactically correct programs throughout the evolution. Grammars are widely used due to their flexibility as they can be defined outside of the GP system to represent the search space of a wide range of problems including program synthesis [20], evolving music [17], managing traffic systems [31] evolving aircraft models [3] and scheduling wireless communications [18,27–30]. Grammar-Guided

Genetic Programming is a variant of GP that use grammar as the representation with most famous variants are Context-Free Grammar Genetic Programming (CFG-GP) by Whigham [33] and grammatical evolution [21].

The G3P system proposed in [7] puts forward a composite and self-adaptive grammar to address different synthesis problems, which solved the limitation of grammar that has to be tailored/adapted for each problem. In [7], several small grammars are defined–each for a data type that defines the function/program to be evolved. Therefore, G3P is able to reuse these grammars for different problems while keeping the search space small by not including unnecessary data types.

2.3 Problem Text Description To/From Source Code

The ability to automatically obtain source code from textual problem descriptions or explain concisely what a block of code is doing have challenged the software engineering community for decades.

The former (i.e., source code from textual description) was aimed at automating the software engineering process with a field mostly divided into two parts: (i) Program Sketching which attempts to lay/generate the general code structure and let either engineers or automated program generative approaches fill the gaps (e.g., [14]), and (ii) Code Retrieval which seeks to find code snippets that highly match the textual description of the problem in large code repositories.

The latter (i.e., textual description from source code) was mostly to increase the readability of source code and assist software engineering with their debugging, refactoring, and porting tasks. Several works have attempted to either provide meaningful comments for specific lines/blocks (e.g., [13]) or to generate brief summaries for the source code (e.g., [1]).

3 Similarity-Based G3P

In this section, we report on how similarity-based G3P perform on selected three problems from [10]. The G3P system in [7] uses error rate fitness function based on given input and output data for evolving the next generation, while similarity-based G3P presented in this section uses code similarity value to the given correct program.

3.1 Proposed Approach

Our ultimate goal is to exploit textual descriptions of user intent in the program synthesis process in combination with current advances in code retrieval/generation (even if such techniques potentially generate multiple incomplete or not fully fit for purpose programs) to guide the search process of G3P. To this end, in this work, we devise a similarity-based G3P system, which uses code similarity to evaluate the fitness of evolved programs against a target source code instead of input/output error rate. The focus of this particular work is not on generating target source code but on (i) assessing the capability of G3P to evolve programs using similarity measures and (ii) identifying the most suitable measure.

3.2 Program Similarity Assessment Approaches

Measuring similarity between source code is a fundamental activity in software engineering. It has multiple applications including duplicate/clone code location, plagiarism detection, code search, security bugs scanning, vulnerability/bugs identification [9] and code recommendation [12]. There have been proposed dozens of similarity detection algorithms since the last few decades, which can be classified into metrics, text, token, tree, and graph-based approaches based on the representation [23]. We selected four top-ranked similarity measures to evaluate their code synthesis proneness when used within G3P.

Cosine Similarity. In addition to the standard code similarity detector, we also used cosine similarity to measure the similarity between two source codes. The following steps illustrate how we measured similarity using cosine similarity:

1. Prepossessing: The source program is tokenized by removing indentation information, including white spaces, brackets, newline characters, and other formatting symbols. Arithmetic operators and assignment symbols were kept as they can provide meaningful structural information.
2. Frequency Computation: For each token sequence of the source program, we compute the frequency of each token.
3. Cosine Similarity Computation: We calculate the similarity score with the cosine formula based on the token frequencies of each source code.

FuzzyWuzzy. [5] is a string matching open-source python library based on difflib python library. It uses Levenshtein Distance to calculate the differences between sequences. The library contains different similarity functions including *TokenSortRatio* and *TokenSetRatio*. Ragkhitwetsagul et al. [23] surprisingly found that the string matching algorithm also works pretty well for measuring code similarity. *TokenSortRatio* function first tokenizes the string by removing punctuation, changing capitals to lowercase. After tokenization, it sorts the tokens alphabetically and then joins them together to calculate the matching score. While *TokenSetRatio* takes out the common tokens instead of sorting them.

CCFinder. [15] is a token-based clone detection technique designed for large-scale source code. The technique detects the code clone with four steps:

1. Lexical Analysis: Generates token sequences from the input source code files by applying a lexical rule of a particular programming language. All source files are tokenized into a single sequence to detect the code clone with multiple files. White spaces and new line characters are removed to detect clone codes with different indentation rules but with the same meaning.

2. Transformation: The system applies transformation rules on token sequence to format the program into a regular structure, allowing it to identify code clones even in codes written with different expressions. Furthermore, all identifiers (e.g., variables, constants, and types) are replaced with special symbols to detect clones with different variable names and expressions.
3. Clone Matching: The suffix-tree matching algorithm is used to compute the matching of the code clones.
4. Formatting: Each clone pair is reported with line information in the source file. This step also contains reformatting from the token sequence.

CCFinder was designed for large-scale programs. Since the codes involved in our evaluation are elementary, the following modifications and simplifications are made to the original tool:

- Given that we are only interested in obtaining a similarity score between two pieces of code, we divide the length of the code clone by the maximum between the lengths of the source files:

$$Sim(x, y) = \frac{Len(Clone(x, y))}{Max(Len(x), Len(y))} \qquad (1)$$

where $Clone(x, y)$ denotes the longest code clone between x and y.
- The matching of code clones using the suffix-tree matching algorithm is simplified by getting the length of the longest common token sequence using a 2D matrix (each dimension representing the token sequence).
- The mapping information between the token sequence and the source code is removed since reporting the line number is no longer needed in our study.

SIM. [8] is a software tool for measuring the structural similarity between two C programs to detect plagiarism in the assignment for lower-level computer science courses. It is also a token-based plagiarism detection tool that uses a string alignment technique to measure code similarity.

The approach comprises two main functions, generating tokens with formatting and calculating the similarity score using alignment. Each source file is first passed through a lexical analyzer to generate a token sequence. Like the common plagiarism detection system, the source code is formatted to standard tokens with white space removal representing arithmetic or logical operators, different symbols, constant or identifiers. After tokenization, the token sequence of the second program is divided into multiple sections, each representing a piece of the original program. These sections are then aligned with the token sequence of the first source code separately, which allows the tool to detect the similarity even the program is plagiarised by modifying the order of the functions.

4   Experiment Setup

4.1   General Program Synthesis Benchmark Suite

Helmuth and Spector [10,11] introduced a set of program synthesis problems. These problems were based on coding problems that might be found in introductory computer science courses. Helmuth and Spector provide a textual description as well as two sets of input/output pairs for both training and testing during the program synthesis process. Table 1 describes the characteristics of each of the program synthesis problems considered in our evaluation.

Table 1. Description and characteristics of the selected program synthesis problems

Problem	Textual Description	# Input/Output Pair	
		Training	Testing
Number IO	Given an integer and a float, print their sum	25	1000
Median	Given 3 integers, print their median	100	1000
Smallest	Given 4 integers, print the smallest of them	100	1000

4.2   Target Programs

To evolve our programs through G3P, we consider an oracle that computes the similarity measure of each evolved program to a target program code obtained using some text to code transformation. In this work, we wish to focus our analysis on the similarity measures and reduce the varying elements in our experiments (particularly in terms of ability to obtain a target program of good quality). Therefore, we consider the theoretical case where the oracle is aware of a code that solves the problem, but it is only reporting the similarity of the evolved code to it. While this assumption is not applicable in real life (i.e., if we know the correct code, then the problem is already solved without requiring any evolution), we hope to get enough insight from it on the capability of G3P to reproduce a program only based on a similarity measure.

Listings 1.1, 1.2, and 1.3 depict the target programs for the oracle assessment of program similarity for Number IO, Smallest, and Median respectively.

```
1  def numberIO(int1, float1):
2      result = float(int1 + float1)
3      return result
```

Listing 1.1. Target program for Number IO

```
1  def smallest(int1, int2, int2, int3):
2      result = min(int1,min(int2,min(int3,int4)))
3      return result
```

Listing 1.2. Target program for Smallest

```
1  def median(int1, int2, int3):
2      if int1 > int2:
3          if int1 < int3:
4              median = int1
5          elif int2 > int3:
6              median = int2
7          else:
```

```
 8              median = int3
 9      else:
10          if int1 > int3:
11              median = int1
12          elif int2 < int3:
13              median = int2
14          else:
15              median = int3
16      return median
```

Listing 1.3. Target program for Median

4.3 G3P Parameter Settings

In our evaluation, we use the same parameter settings as those defined for G3P [7]. We only introduce a unique varying element (i.e., the fitness function based on a particular similarity measure). We repeat our evaluations 30 times for each problem and each G3P version (each version with its specific similarity measure). The general settings for the G3P system are shown in Table 2.

Table 2. Experiment parameter settings

Parameter	Setting	Parameter	Setting
Runs	30	Mutation probability	0.05
Generation	200	Node limit	250
Population size	1000	Variable per type	3
Tournament size	7	Max execution time	1 s
Crossover probability	0.9		

5 Results

In this section, we report and discuss the results of our evaluations. First, we start by comparing the performance of G3P using each of the similarity measures, then we compare them against the traditional error-rate based G3P.

5.1 Comparison of Similarity Measures

The result of the similarity-based G3P is reported in this subsection. The goal of this experiment is to assess G3P's ability to evolve a program solving a program synthesis problem (only known to an oracle) based on the similarity measure.

Figure 1 shows the number of runs (out of 30) where G3P manages to evolve the correct program for each of the program synthesis problems while using one of the four considered similarity measures as the fitness function.

We see from Fig. 1 that G3P was able to able to evolve the correct programs for Number IO and Smallest at least once with Cosine, CCFinder and SIM. However, G3P did not manage to evolve any correct program for Median. G3P manages to find the correct program for Number IO in most runs (i.e., 27 out of 30) while using Cosine Similarity. However, the same program fails to find any correct program for Smallest. Similarly, G3P manages to find the correct program with Smallest in 17 runs out of 30 while using SIM, but the same program fails to find any correct program for Number IO. Alternatively G3P with CCFinder finds correct programs for both Number IO and Smallest, but in fewer runs. Overall, we could say that G3P has the potential to evolve programs for synthesis problems using similarity measures. However, there is no similarity measure that seems to work better than the rest and we need to consider combining similarity measures to increase the effectiveness of the approach.

5.2   Comparison Against Error Rate-Based G3P

While we have seen that G3P has the capability to evolve correct programs for some program synthesis, we would like to assess how efficient is this process at evolving correct programs in comparison with the use of input/output error rate.

Figure 2 shows the performance of G3P with the input/output error rate to evolve correct programs for each of the considered programs over 30 distinct runs. We see that G3P with input/output error rate is capable to evolve correct problems to all the considered program synthesis problems. Furthermore, it is also capable of finding a correct program in more runs than the different G3P approaches using any similarity measure. Therefore, while we have seen that similarity measures seem promising to guide the G3P search for correct programs to program synthesis problems, they are not reaching the performance level of the traditional input/output error rate. This difference could be explained by the long amount of research that has been carried out to refine and optimise the G3P process with input/output error rate (particularly in terms of designing fit for purpose crossover and mutation operators). Therefore, one potential approach could be to use both the similarity objectives and the error-rate to guide the evolutionary process in a multi-objective approach [32].

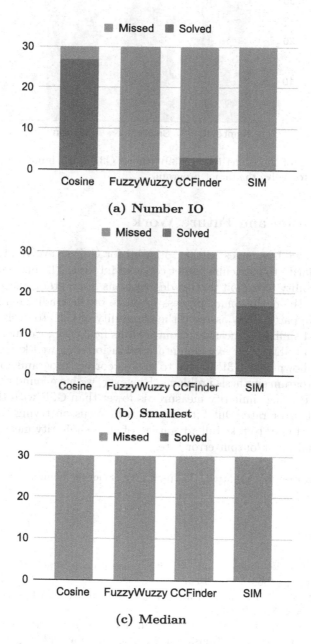

Fig. 1. Number of iterations (out of 30) where G3P manages or fails to evolve the target program with each of the similarity measures.

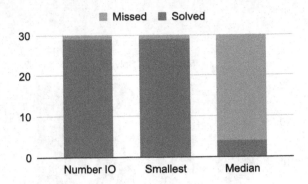

Fig. 2. Number of iterations (out of 30) where G3P with input/output error rate manages/fails to evolve the target program.

6   Conclusion and Future Work

In this paper, we have assessed the potential for G3P to evolve programs based on their similarity to particular target codes of interest. The ultimate goal of this work is the ability to exploit textual descriptions of program synthesis problems as a guide to the evolutionary process in place of the traditional input/output error rate. We particularly assessed the capability of G3P to evolve correct programs using 4 similarity measures from various fields (i.e., Cosine, FuzzyWuzzy, CCFinder, and SIM). Our experimental evaluation on a well-known benchmark dataset has shown that G3P is able to evolve correct programs for some of the considered program synthesis problems. However, we have found that the performance of G3P using similarity measures is lower than G3P with the traditional input/output error rate. Out future work will focus on trying to improve the performance of G3P to take full advantage of such similarity measures alongside the traditional input/output error rate.

Acknowledgement. Supported, in part, by Science Foundation Ireland grant 13/RC/2094_P2.

References

1. Alexandru, C.V.: Guided code synthesis using deep neural networks. In: ACM SIGSOFT, pp. 1068–1070 (2016)
2. Brameier, M., Banzhaf, W., Banzhaf, W.: Linear Genetic Programming, vol. 1. Springer, New York (2007)
3. Byrne, J., Cardiff, P., Brabazon, A., et al.: Evolving parametric aircraft models for design exploration and optimisation. Neurocomputing **142**, 39–47 (2014)
4. Ciritoglu, H.E., Saber, T., Buda, T.S., Murphy, J., Thorpe, C.: Towards a better replica management for hadoop distributed file system. In: IEEE BigData Congress (2018)
5. Cohen, A.: Fuzzywuzzy: fuzzy string matching in python (2011)

6. Forstenlechner, S.: Program synthesis with grammars and semantics in genetic programming. Ph. D. dissertation (2019)
7. Forstenlechner, S., Fagan, D., Nicolau, M., O'Neill, M.: A grammar design pattern for arbitrary program synthesis problems in genetic programming. In: McDermott, J., Castelli, M., Sekanina, L., Haasdijk, E., García-Sánchez, P. (eds.) EuroGP 2017. LNCS, vol. 10196, pp. 262–277. Springer, Cham (2017). https://doi.org/10.1007/978-3-319-55696-3_17
8. Gitchell, D., Tran, N.: Sim: a utility for detecting similarity in computer programs. ACM SIGCSE Bull. **31**(1), 266–270 (1999)
9. Hartmann, B., MacDougall, D., Brandt, J., Klemmer, S.R.: What would other programmers do: suggesting solutions to error messages. In: SIGCHI, pp. 1019–1028 (2010)
10. Helmuth, T., Spector, L.: Detailed problem descriptions for general program synthesis benchmark suite. University of Massachusetts Amherst (2015)
11. Helmuth, T., Spector, L.: General program synthesis benchmark suite. In: GECCO, pp. 1039–1046 (2015)
12. Holmes, R., Murphy, G.C.: Using structural context to recommend source code examples. In: ICSE, pp. 117–125 (2005)
13. Hu, X., Li, G., Xia, X., Lo, D., Jin, Z.: Deep code comment generation. In: IEEE/ACM ICPC, pp. 200–20010 (2018)
14. Jeon, J., Qiu, X., Foster, J.S., Solar-Lezama, A.: Jsketch: sketching for java. In: ESEC/FSE, pp. 934–937 (2015)
15. Kamiya, T., Kusumoto, S., Inoue, K.: Ccfinder: A multilinguistic token-based code clone detection system for large scale source code. IEEE Trans. Softw. Eng. **28**(7), 654–670 (2002)
16. Koza, J.R., et al.: Genetic Programming II, vol. 17. MIT Press, Cambridge (1994)
17. Loughran, R., McDermott, J., O'Neill, M.: Tonality driven piano compositions with grammatical evolution. In: IEEE CEC, pp. 2168–2175 (2015)
18. Lynch, D., Saber, T., Kucera, S., Claussen, H., O'Neill, M.: Evolutionary learning of link allocation algorithms for 5G heterogeneous wireless communications networks. In: GECCO, pp. 1258–1265 (2019)
19. Miller, J.F., Harding, S.L.: Cartesian genetic programming. In: GECCO, pp. 2701–2726 (2008)
20. O'Neill, M., Nicolau, M., Agapitos, A.: Experiments in program synthesis with grammatical evolution: a focus on integer sorting. In: CEC, pp. 1504–1511 (2014)
21. O'Neill, M., Ryan, C.: Grammatical Evolution: Evolutionary Automatic Programming in a Arbitrary Language, vol. 4 of Genetic Programming (2003)
22. Pantridge, E., Spector, L.: Pyshgp: pushgp in python. In: GECCO, pp. 1255–1262 (2017)
23. Ragkhitwetsagul, C., Krinke, J., Clark, D.: A comparison of code similarity analysers. Empir. Softw. Eng. **23**(4), 2464–2519 (2018). https://doi.org/10.1007/s10664-017-9564-7
24. Saber, T., Brevet, D., Botterweck, G., Ventresque, A.: Is seeding a good strategy in multi-objective feature selection when feature models evolve? IST (2017)
25. Saber, T., Brevet, D., Botterweck, G., Ventresque, A.: MILPIBEA: algorithm for multi-objective features selection in (evolving) software product lines. In: Paquete, L., Zarges, C. (eds.) EvoCOP 2020. LNCS, vol. 12102, pp. 164–179. Springer, Cham (2020). https://doi.org/10.1007/978-3-030-43680-3_11
26. Saber, T., Delavernhe, F., Papadakis, M., O'Neill, M., Ventresque, A.: A hybrid algorithm for multi-objective test case selection. In: IEEE CEC (2018)

27. Saber, T., Fagan, D., Lynch, D., Kucera, S., Claussen, H., O'Neill, M.: A hierarchical approach to grammar-guided genetic programming: the case of scheduling in heterogeneous networks. In: Fagan, D., Martín-Vide, C., O'Neill, M., Vega-Rodríguez, M.A. (eds.) TPNC 2018. LNCS, vol. 11324, pp. 225–237. Springer, Cham (2018). https://doi.org/10.1007/978-3-030-04070-3_18

28. Saber, T., Fagan, D., Lynch, D., Kucera, S., Claussen, H., O'Neill, M.: Multi-level grammar genetic programming for scheduling in heterogeneous networks. In: Castelli, M., Sekanina, L., Zhang, M., Cagnoni, S., García-Sánchez, P. (eds.) EuroGP 2018. LNCS, vol. 10781, pp. 118–134. Springer, Cham (2018). https://doi.org/10.1007/978-3-319-77553-1_8

29. Saber, T., Fagan, D., Lynch, D., Kucera, S., Claussen, H., O'Neill, M.: Hierarchical grammar-guided genetic programming techniques for scheduling in heterogeneous networks. In: CEC (2020)

30. Saber, T., Fagan, D., Lynch, D., Kucera, S., Claussen, H., O'Neill, M.: A multi-level grammar approach to grammar-guided genetic programming: the case of scheduling in heterogeneous networks. Genet. Program. Evolvable Mach. 20(2), 245–283 (2019). https://doi.org/10.1007/s10710-019-09346-4

31. Saber, T., Wang, S.: Evolving better rerouting surrogate travel costs with grammar-guided genetic programming. In: IEEE CEC, pp. 1–8 (2020)

32. Tao, N., Ventresque, A., Saber, T.: Multi-objective grammar-guided genetic programming with code similarity measurement for program synthesis. In: IEEE CEC (2022)

33. Whigham, P.A.: Grammatical bias for evolutionary learning (1997)

Author Index

Printed in the United States
by Baker & Taylor Publisher Services